In Clinical Practice

Taking a practical approach to clinical medicine, this series of smaller reference books is designed for the trainee physician, primary care physician, nurse practitioner and other general medical professionals to understand each topic covered. The coverage is comprehensive but concise and is designed to act as a primary reference tool for subjects across the field of medicine.

Theodore Dassios

Clinical Respiratory Physiology of the Newborn

Theodore Dassios [iD]
Women & Children's Health Department
Kings College London
London, UK

Neonatal Intensive Care Unit
University of Patras
Patras, Greece

ISSN 2199-6652 ISSN 2199-6660 (electronic)
In Clinical Practice
ISBN 978-3-032-05737-2 ISBN 978-3-032-05738-9 (eBook)
https://doi.org/10.1007/978-3-032-05738-9

© The Editor(s) (if applicable) and The Author(s), under exclusive license to Springer Nature Switzerland AG 2025

This work is subject to copyright. All rights are solely and exclusively licensed by the Publisher, whether the whole or part of the material is concerned, specifically the rights of translation, reprinting, reuse of illustrations, recitation, broadcasting, reproduction on microfilms or in any other physical way, and transmission or information storage and retrieval, electronic adaptation, computer software, or by similar or dissimilar methodology now known or hereafter developed.
The use of general descriptive names, registered names, trademarks, service marks, etc. in this publication does not imply, even in the absence of a specific statement, that such names are exempt from the relevant protective laws and regulations and therefore free for general use.
The publisher, the authors and the editors are safe to assume that the advice and information in this book are believed to be true and accurate at the date of publication. Neither the publisher nor the authors or the editors give a warranty, expressed or implied, with respect to the material contained herein or for any errors or omissions that may have been made. The publisher remains neutral with regard to jurisdictional claims in published maps and institutional affiliations.

This Springer imprint is published by the registered company Springer Nature Switzerland AG
The registered company address is: Gewerbestrasse 11, 6330 Cham, Switzerland

If disposing of this product, please recycle the paper.

Preface

The content of this book is broadly based on a series of lectures on this topic which I have delivered to the postgraduate students in King's College London and the neonatal trainees at King's College Hospital. I have tried to refine the content over the years, so that what is conveyed can be better understood in the clinical setting, and have backed up the physiological background with real-life clinical scenarios where this knowledge can be directly applied.

This book has a simple main aim: to introduce respiratory function and physiology to neonatal professionals in a way that can help them in their everyday clinical practice. While in recent years major technological advances have taken place in neonatal respiratory medicine, with ever new and multi-potent ventilators and monitoring systems, basic understanding of the physiological basis of disease might still constitute the strongest bedside ally of the clinician. Irrespective of how complicated a system might develop to be, solid understanding of the principles of how the respiratory system works in health and in disease will remain our greatest friend in times of need.

It is not my intention to fully describe all the physiological and pathophysiological phenomena which can be observed in the respiratory system of the newborn, especially as a single author, as there is undoubtedly a plethora of very knowledgeable researchers and clinicians in this field around the world, many of whom I have had the pleasure to meet and work with. Practically, the aim of this book is that if a neonatal trainee or other healthcare professional has read this information and is struggling to manage a difficult baby who cannot oxygenate at four in the morning, some of the principles described in this book can maybe get them through the night and, possibly at a later time, get them interested to expand on the understanding of these general principles. I hope that all students of this medical field, either at the undergraduate or the postgraduate level will also find the information in this book useful.

It is generally accepted that in our rapidly changing world and working environment, artificial intelligence systems will outperform humans in diagnostic tasks, especially the ones that involve image analysis, differential diagnosis and pattern recognition. In medicine however, some human qualities might prove difficult to overcome, such as empathy, hand dexterity, creativity, ethical decision making and the ability to synthesise information at varying hierarchical degrees of importance. For these, I think we will still need a well-read clinician.

Since my days in training, I was faced with what I saw as a controversy. Some more senior neonatologists were certain in stating that "lung research is over, it has all been done by now". However, if one went to a congress, it was the lung sessions that were among the best attended. More and more interventions and technological approaches became available and integrated in clinical care, often on a yearly basis. People wanted to know both the advanced and the basic principles that underpin these advances. Whenever we organised a course on invasive ventilation or non-invasive support as part of a congress programme, it was always oversubscribed, and we had to close registration early. I came to understand that people are genuinely interested in this topic. This book is one way to try and meet this demand.

As we move into the future, it is useful to also look back and remember the Hippocratic saying: *"Wherever the art of medicine is loved, there is also a love of humanity."*

London, UK Theodore Dassios

Competing Interests The author has no competing interests to declare that are relevant to the content of this manuscript.

Contents

1 The Respiratory System Before and After Birth ... 1
   Foetal and Postnatal Development 2
   Surfactant and Foetal Lung Fluid 2
   Respiratory Transition at Birth 3
   Postnatal Lung Growth 5
   The Structure-Function Relationship
   in the Human Lungs 7
   Questions.................................... 8
   References................................... 9

2 Common Neonatal Respiratory Diseases 11
   Respiratory Distress Syndrome................... 11
   Bronchopulmonary Dysplasia.................... 12
   Term Infants 12
   Congenital Diseases 13
   Other Respiratory Complications 13
   Questions.................................... 14
   References................................... 14

3 Ventilation 17
   Tidal Volume Compartments..................... 18
   Volume Targeted Ventilation.................... 22
   Calculation of Dead Spaces 23
   Alternative Mechanisms of Gas Transport 28
   Questions.................................... 29
   References................................... 30

4	**Diffusion**.	33
	Alveolar-Arterial Gradient	34
	Decreased Surface Area and Thickened Membrane	36
	Questions.	38
	References.	38
5	**Perfusion**	41
	Pulmonary Vascular Resistance	42
	"Fixed" Pulmonary Hypertension: Congenital Diaphragmatic Hernia—Bronchopulmonary Dysplasia.	45
	Intrapulmonary Shunting	48
	Ductus Arteriosus	49
	Vascular Origins of Respiratory Disease	50
	Questions.	50
	References.	51
6	**Ventilation to Perfusion relationships**	55
	Ventilation to Perfusion Mismatch	56
	Measuring the Ventilation to Perfusion Ratio	58
	Interventions to Improve V_A/Q Matching	60
	Questions.	61
	References.	61
7	**Oxygen Transport to the Tissues**	63
	The Oxyhaemoglobin Dissociation Curve	64
	Haemoglobin Subtypes.	68
	Carbon Dioxide.	70
	Questions.	70
	References.	71
8	**Mechanics of Breathing**	73
	Compliance.	73
	Compliance Sigmoid Curve	75
	Lung Recruitment Manoeuvres.	77
	Resistance.	78
	Increased Airway Resistance	80
	Flow-Volume Curves	81
	Inspiratory Time constant	83

	Questions.	85
	References.	87
9	**Work of Breathing**	89
	The Neonatal Diaphragm	90
	Functional Properties of the Respiratory Muscles	91
	Methods to Measure Neonatal Respiratory Muscle Function	91
	Prediction of Extubation	93
	Factors That Affect Respiratory Muscle Function in the Newborn	94
	Neurally Adjusted Ventilatory Assist	96
	Questions.	98
	References.	98
10	**Control of Respiration**	103
	Question	104
	References.	105
11	**Waveforms**	107
	Pressure Versus Time	109
	Flow Versus Time	111
	Volume Versus Time	114
	Pressure Versus Volume	115
	Flow Volume Curve.	120
	Carbon Dioxide Versus Time	122
	Rebreathing.	126
	Carbon Dioxide Versus Volume	128
	Questions.	130
	References.	132
12	**The Neonatal Respiratory System at Critical Extremes**	135
	Periviable Gestation	135
	Congenital Diaphragmatic Hernia	137
	Extubation.	138
	Altitude Hypoxia.	139
	Questions.	142
	References.	143

13 Children and Adults with a History of Neonatal Lung Disease ... 149
Known Unknowns ... 149
Unknown Unknowns ... 150
The Respiratory BPD Phenotype in Older Children and Adults ... 152
- Obstructive Lung Disease ... 153
- Hyperinflation ... 154
- Restrictive Lung Disease ... 154
- Exercise Capacity ... 154
- Pulmonary Hypertension ... 155
- Dysynaptic Lung Growth ... 155

Questions ... 155
References ... 156

14 Lung Function Tests in Neonates ... 159
Question ... 163
References ... 163

Correct Answers to Chapter Questions ... 167

The Respiratory System Before and After Birth

The main function of the respiratory system is gas exchange in the form of oxygen intake and carbon dioxide removal. For gas exchange to take place, gas will need to be transported to the alveolar membrane which is the sum of approximately 500 million alveoli in a healthy adult and forms a very thin and large surface where the atmospheric gas meets the pulmonary circulation. The transfer of the respiratory gases to the alveolar membrane requires the activation of the respiratory muscles, which will need to contract and generate negative intrathoracic pressure to draw air into the lungs from the surrounding environment.

Gas exchange is a passive phenomenon, where a gas diffuses through the alveolar membrane from a compartment of a higher concentration towards one with a lower concentration. The gases are then delivered via the pulmonary and the systemic circulation to the peripheral tissues with the help of the haemoglobin which is the main oxygen-binding protein.

The neonatal respiratory system has some unique functional and anatomical characteristics relating to the intrauterine stages of lung development and the transition of oxygenation from the intrauterine gas exchange interface, which is the placenta, to the newborn lungs immediately after birth. These characteristics possibly explain why the incidence of respiratory disease is increased in the newborn, and why neonatal respiratory disease is frequently developmental in origin and is associated with life-long consequences for respiratory and overall health.

Foetal and Postnatal Development

The intrauterine development of the respiratory system is divided into five phases [1]. During the embryonic phase (until the eighth week of gestation) the two primary lung buds are formed, stemming from the respiratory diverticulum which originates from the anterior foregut. The five lobar buds which will later form the three right and the two left lung lobes first appear close to day 30 after conception. In the pseudoglandular phase (5th–17th week) the airways will undergo repeated consecutive divisions and the foetal lungs will have 16–25 generations of primitive airways. Although the existing airways will further elongate and widen, no new airways will develop after this stage. During the canalicular phase (16th–26th week) epithelialised airways will grow rapidly and penetrate the mesenchymal tissue. The epithelial cells will also differentiate to type-I and type-II pneumocytes which are responsible for forming the blood-air barrier and produce surfactant respectively. It is only at the end of this stage that some limited gas exchange, and thus viability, can be achieved. In the saccular phase (24th–36th week of gestation) the airspaces will increase in size, while the walls of the airspaces will become thinner. The alveolar phase (36th week of gestation-birth) corresponds to the development of the alveoli which occurs by division of the primary septa to secondary ones, which will then form the alveolar membrane [1]. The layers of the alveolar membrane are thus: the epithelial wall of the type I pneumocytes, the basal membranes of the epithelial and the endothelial cells and the endothelial cells.

Surfactant and Foetal Lung Fluid

Pulmonary surfactant is a lipoprotein consisting of phospholipid, cholesterol and surfactant proteins and is produced by the type-II pneumocytes [2]. The biological action of surfactant is to reduce the surface tension in the lungs and thus prevent alveolar collapse and atelectasis. The synthesis of pulmonary surfactant develops in the second half of gestation and continues to increase until birth. Absence of adequate endogenous surfactant constitutes a major aetiological factor of respiratory disease in the newborn and especially in the ones born prematurely. During the intrauterine period, the foetal lung cells secrete a fluid, which fills the lungs and contributes to lung development [3]. The foetal lung fluid promotes lung growth by distending the developing lung. This mechanism has found a therapeutic application in an antenatal intrauterine intervention called foetoscopic endoluminal tracheal occlusion (FETO), which is offered to infants with congenital diaphragmatic hernia [4]. In those infants, the postulated mechanism is that tracheal occlusion will entrap the foetal lung fluid within the developing lungs, which will increase the distending pressure and thus promote lung growth. Foetal breathing movements, combined with adequate foetal lung liquid, contribute to lung development and maturation [5].

Respiratory Transition at Birth

During intrauterine life, the source of oxygenation is the placenta. The lungs in utero are not performing gas exchange and are partially bypassed by the right circulation as most of the content of the pulmonary circulation is shunted into the left circulation via the foramen ovale and the ductus arteriosus (Fig. 1.1). The pulmonary vessels are thus functionally dormant in utero as they accommodate a smaller blood volume while they also have a high vascular resistance. It is interesting to remember that during foetal life, due to shunting of deoxygenated blood into the left circulation, the foetal peripheral oxygen saturation remains in the lower range of 60% but increases rapidly to above 90% within the first ten minutes after birth [6].

At birth, and following the first breaths, the newborns clear the foetal lung fluid through absorption via the pulmonary lymphatics while a part of the foetal lung fluid is also ejected via the upper airways. Lungs are expanded with air, and a functional residual capacity is established. As the umbilical cord is clamped or pulsa-

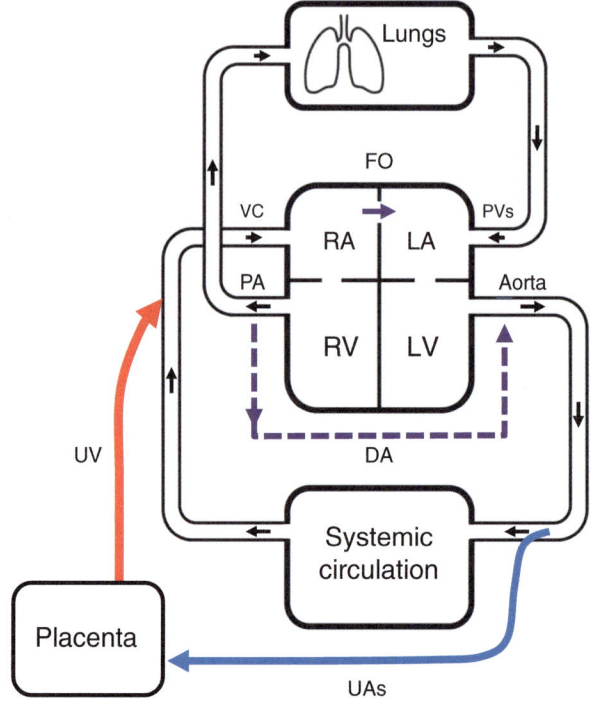

Fig. 1.1 Schematic depiction of the foetal circulation. Oxygenated blood from the placenta is delivered to the right circulation via the umbilical vein (UV). Shunting of the oxygen-rich blood from the placenta occurs at the level of the foramen ovale (FO) and the ductus arteriosus (DA). Deoxygenated blood is returned to the placenta via the umbilical arteries (UAs). RA right atrium, RV right ventricle, LA left atrium, LV left ventricle, VC vena cavae, PA pulmonary artery, PVs pulmonary veins

tion stops following separation of the placenta from the uterine wall, the systemic pressure rises and the foramen ovale closes acting as a flap valve (one-way valve). The clamping of the cord also leads to a cessation of blood flow through the umbilical vein which then causes the ductus arteriosus to collapse. The relative increase in oxygen tension after birth, leads to vasodilation of the pulmonary vasculature and a decrease in the pulmonary artery pressure [1]. The combined effect of this increase in the systemic circulation pressure, concurrently with a decrease in the pulmonary pressure, leads to a decrease of the right-to-left flow through the ductus arteriosus.

Postnatal Lung Growth

As mentioned above, the process of alveolarisation (the production of new alveoli) starts in the middle of foetal life in such a way that at birth, a term infant has approximately 150 million alveoli [7]. This process continues postnatally to achieve the final number of 400–600 million which is reached later in childhood [8]. Similarly, the volume of the lungs also increases throughout childhood until approximately 25 years of age. Geometrically and at the microscopic level, the alveoli as they develop and increase in number attain a polyhedral shape in the three dimensions, more usually that of a dodecahedron, which at a planar histological level would appear as polygonal and most commonly pentagonal (Fig. 1.2) [9].

Lung function measured over the lifespan at a population level, seems to follow a predefined trajectory (lung function tracking) which in health is predictable to some extent, from early childhood to adulthood and during aging (Fig. 1.3) [10]. Early abnormalities in the neonatal or foetal life such as prematurity and intrauterine growth retardation can assign individuals to a lower tracking trajectory which can be further exacerbated by postnatal factors such as exposure to tobacco or air pollution, early viral infections, development of asthma or atopy and potentially overnutrition or exposure to abnormal microbiomes [11].

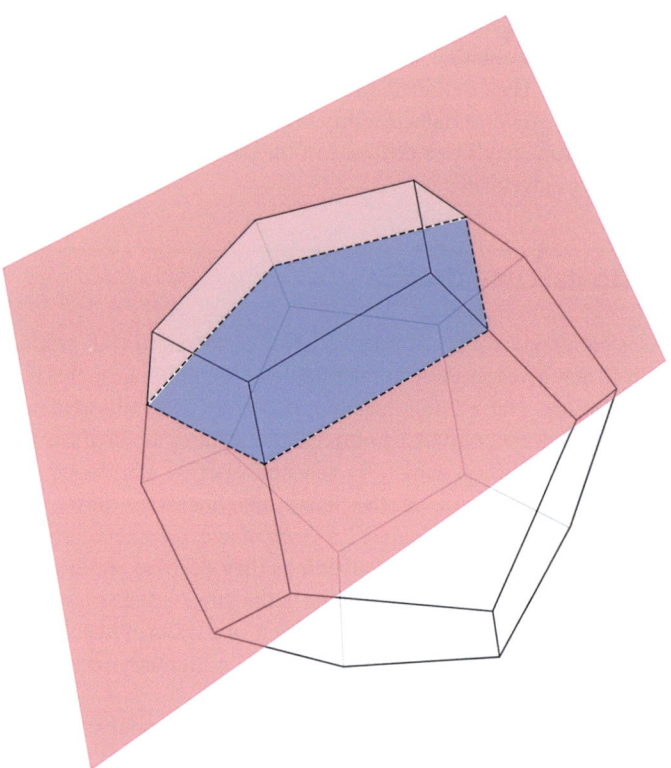

Fig. 1.2 Geometrical microscopical appearance of the alveoli. A solid canonical dodecahedron with a planar intersection resulting in a shape of a pentagon

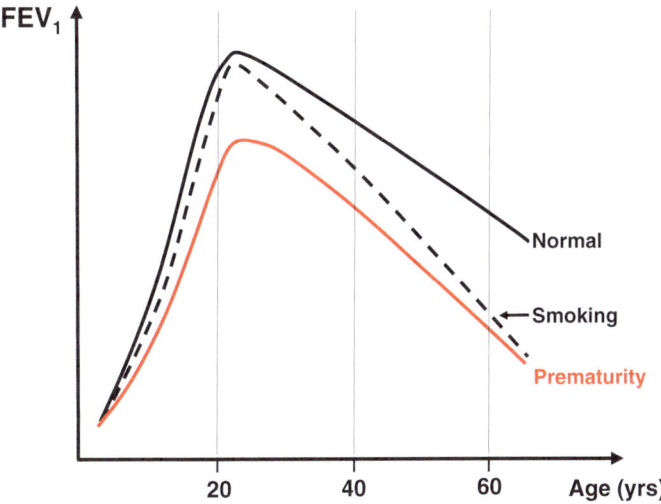

Fig. 1.3 Lung function trajectories. The normal curve follows a more gradual decline than the curve corresponding to smoking. The curve of preterm infants does not reach the maximum value compared to the normal curve and potentially has a faster rate of decline

The Structure-Function Relationship in the Human Lungs

The structure of the human respiratory system is explained by the functional needs it serves, mainly the requirement to provide a large and thin interface for gas exchange. The surface of the alveolar membrane is approximately 50–100 m^2 in a healthy adult, and this large surface is contained in a relatively small volume of approximately six litres, which is the expected value for the total lung capacity. This sophisticated packing mechanism serves the need to accommodate the oxygen needs at rest and during increased demands such as during strenuous aerobic activity [12]. This function is achieved by repeated airway divisions which start from the trachea and continue down to level of the alveoli. These consecutive subdivisions finally generate approximately 500 mil-

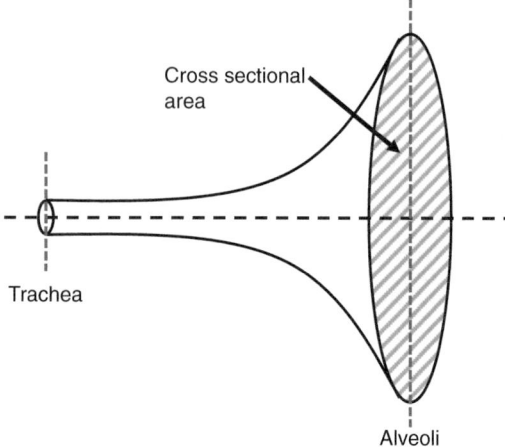

Fig. 1.4 The trumpet model. The total cross sectional area of the airways increases from the trachea towards the terminal bronchioles

lion alveoli in a fully developed human lung [13]. As discussed later in the chapter on pulmonary mechanics and airways resistance, the diameter of each airway decreases with every division (for example the diameter of the trachea is larger than the diameter of the left or right main bronchus), but the total cross sectional area of the airways increases from the trachea to the terminal bronchioles in a way that schematically resembles a trumpet bell (Fig. 1.4). This structure is associated with a decreasing resistance to airway flow towards the later divisions of the bronchial tree and facilitates the transfer and diffusion of gases.

Questions

Question 1: The following are true for the foetal lung fluid:
  (a) Is secreted by the foetal lung cells
  (b) Is amniotic fluid
  (c) Promotes foetal lung development
  (d) Is entrapped with foetal endoluminal tracheal occlusion

Question 2: The following is true concerning the foetal circulation during intrauterine life:
 (a) The pulmonary vessels have a low vascular resistance
 (b) The pulmonary vein transfers deoxygenated blood
 (c) The foramen ovale can shunt blood both ways (right to left and left to right)
 (d) Foetal arterial saturation is above 90%
 (e) The placenta is the main source of oxygen

Question 3: In postnatal life:
 (a) There is no further increase in the number of the alveoli
 (b) There is further increase in the number of airways
 (c) Postnatal lung function can be determined by antenatal events
 (d) It is normal for the pulmonary vascular resistance to remain high

References

1. Schoenwolf GC, Larsen WJ. Larsen's human embryology, vol. xix. 4th ed. Philadelphia: Churchill Livingstone/Elsevier; 2009. 687 pp.
2. Soll RF, Blanco F. Natural surfactant extract versus synthetic surfactant for neonatal respiratory distress syndrome. Cochrane Database Syst Rev. 2001;(2):CD000144. Epub 2001/06/19.10.1002/14651858.CD000144.
3. Olver RE, Walters DV, Wilson SM. Developmental regulation of lung liquid transport. Annu Rev Physiol. 2004;66:77–101. Epub 2004/02/24.10.1146/annurev.physiol.66.071702.145229.
4. Jani JC, Nicolaides KH, Gratacos E, Vandecruys H, Deprest JA, Group FT. Fetal lung-to-head ratio in the prediction of survival in severe left-sided diaphragmatic hernia treated by fetal endoscopic tracheal occlusion (FETO). Am J Obstet Gynecol. 2006;195(6):1646–50. Epub 2006/06/14.10.1016/j.ajog.2006.04.004.
5. Adzick NS, Harrison MR, Glick PL, Villa RL, Finkbeiner W. Experimental pulmonary hypoplasia and oligohydramnios: relative contributions of lung fluid and fetal breathing movements. J Pediatr Surg. 1984;19(6):658–65. Epub 1984/12/01.10.1016/s0022-3468(84)80349-8.
6. Dildy GA, van den Berg PP, Katz M, Clark SL, Jongsma HW, Nijhuis JG, et al. Intrapartum fetal pulse oximetry: fetal oxygen saturation trends during labor and relation to delivery outcome. Am J Obstet Gynecol. 1994;171(3):679–84. Epub 1994/09/01.10.1016/0002-9378(94)90081-7.

7. Hislop AA, Wigglesworth JS, Desai R. Alveolar development in the human fetus and infant. Early Human Dev. 1986;13(1):1–11. Epub 1986/02/01.10.1016/0378-3782(86)90092-7.
8. Narayanan M, Owers-Bradley J, Beardsmore CS, Mada M, Ball I, Garipov R, et al. Alveolarization continues during childhood and adolescence: new evidence from helium-3 magnetic resonance. Am J Respir Critic Care Med. 2012;185(2):186–91. Epub 2011/11/11.10.1164/rccm.201107-1348OC.
9. Linhartova A, Caldwell W, Anderson AE. A proposed alveolar model for adult human lungs: the regular dodecahedron. Anat Rec. 1986;214(3):266-272. Epub 1986/03/01.10.1002/ar.1092140305
10. Bolton CE, Bush A, Hurst JR, Kotecha S, McGarvey L. Lung consequences in adults born prematurely. Thorax. 2015;70(6):574–80. Epub 2015/04/01.10.1136/thoraxjnl-2014-206590.
11. Bush A. Lung development and aging. Annal Am Thorac Soc. 2016;13(Suppl 5):S438–S46. Epub 2016/12/23.10.1513/AnnalsATS.201602-112AW.
12. West JB. Respiratory physiology : the essentials, vol. viii. 9th ed. Philadelphia: Wolters Kluwer Health/Lippincott Williams & Wilkins; 2012. 200 pp.
13. Weibel ER. Morphometry of the human lung, vol. xi. Berlin: Springer; 1963. 151 pp.

Common Neonatal Respiratory Diseases

2

While this book is generally mostly addressed to neonatal health professionals, it might be useful to summarise the most common respiratory conditions encountered in the newborn period and some key epidemiological and pathophysiological characteristics. This will hopefully be helpful for the non-neonatal reader as these disorders are mentioned throughout the following chapters.

Respiratory Distress Syndrome

Respiratory distress syndrome (RDS) is the most common respiratory disease in the newborn and affects primarily preterm infants. It is more common and more severe in the most premature infants and is primarily due to the lack of natural surfactant in the context of pulmonary immaturity [1]. Exogenous surfactant administration is the main aetiological treatment for RDS, alongside supportive therapy with mechanical ventilation, non-invasive support and provision of supplemental oxygen as required. As a rough estimate for the frequency and prevalence of the disease, approximately half of all infants born between 22 and 33 weeks of gestation will receive endotracheal surfactant, with some milder cases managed only on non-invasive support without surfactant administration, which means that the total prevalence of the syndrome is higher [2].

Bronchopulmonary Dysplasia

The term bronchopulmonary dysplasia (BPD) refers to the *chronic phase* of respiratory disease in prematurely born infants who acutely and in the first days of life suffered from RDS. Although for individual infants BPD is the continuation of RDS falling into a chronic phase, it is not solely due to prematurity and lack of surfactant, but is a multifactorial disease which is aggravated by lung trauma secondary to invasive ventilation, oxygen toxicity and a plethora of perinatal and postnatal factors, including fetal and postnatal growth retardation, inflammation and a persistently open ductus arteriosus which might cause haemodynamic compromise [3].

By convention, BPD is diagnosed in infants born before 32 completed weeks of gestation who needed supplemental oxygen at least for the first 28 days of life. These infants are assessed again at 36 weeks of postconceptional age, and if there is no oxygen or any requirement for respiratory support at that point, the disease is classified as mild, while if they need supplemental oxygen or positive pressure ventilatory support (either invasive or non-invasive) they are then classified as moderate or severe [4]. The main morphological consequence of BPD is a simplified lung structure with larger and fewer alveoli, which is secondary to an arrest or severe disruption of the alveolarisation process [5]. BPD is more common in the most immature infants with 57% of infants born at less than 28 weeks of gestation developing moderate or severe disease in a recent whole population cohort [6].

Term Infants

The incidence of respiratory disease in the term population is lower compared to the preterm ones, but the sheer volume of term infants who are admitted with transient respiratory pathology is considerably larger. Such pathology includes the transient tachypnoea of the newborn (TTN) a rather benign and self-limited condition which predominantly affects late preterm or term infants born via caesarean section and is thought to be related to delayed clearance of the lung fluid possibly in the context of the lack of epinephrine and nor-epinephrine secretion due to the absence of vaginal birth [7]. The acute phase of the condition usually lasts for a few days and has limited impact on future lung function and structure [8].

Term infants can also be affected by meconium aspiration syndrome (MAS) a condition which commonly affects post-term infants (born after their due date) or infants in foetal stress or hypoxia [9]. Meconium is the foetal stool, which although is usually "clean" and free of bacteria, it can by inhaled into the lungs and cause severe pneumonitis and oxygenation failure which is often complicated by acute and persistent pulmonary hypertension. Management of the severe cases of MAS is generally supportive and the disease can be challenging and is associated with increased morbidity and mortality [9].

Congenital Diseases

Congenital diaphragmatic hernia (CDH) is a rare birth defect caused by incomplete development of the diaphragm, which results in herniation of abdominal viscera – most commonly intestine – into the thoracic cavity. The compression of the developing lungs in utero can lead to critical pulmonary parenchymal hypoplasia affecting primarily the ipsilateral but also the contralateral lung with arrested airway development and a decreased number of alveoli. The pulmonary vessels are also severely affected at the macroscopic and microscopic level, with poor vascular responsiveness and increased muscularity and wall thickness [10]. Infants with CDH have high mortality and multi-system morbidity [11] and lung function abnormalities which persist into later life [12].

Other Respiratory Complications

Air leak complications such as pneumothorax, pneumomediastinum and pulmonary interstitial emphysema can occur in newborn infants and are more common in the preterm population and in the group of infants who were invasively ventilated [13]. Although beyond the scope of this narrative, numerous other complications and primary pathologies can arise in the newborn such as pneumonia, pulmonary haemorrhage, pleural effusion, surfactant protein deficiencies, congenital pulmonary airway malformation and others [14].

Questions

Question 1: The following statements are true:
(a) Respiratory distress syndrome can only affect premature infants
(b) Surfactant is the only etiological therapy in RDS
(c) Bronchopulmonary dysplasia is usually diagnosed at 32 weeks postmenstrual age
(d) Meconium aspiration cannot happen in prematurely-born infants
(e) Congenital diaphragmatic hernia affects the pulmonary vasculature

References

1. Sweet DG, Carnielli VP, Greisen G, Hallman M, Klebermass-Schrehof K, Ozek E, et al. European Consensus Guidelines on the management of respiratory distress syndrome: 2022 Update. Neonatology. 2023;120(1):3–23. Epub 2023/03/03.10.1159/000528914.
2. Haumont D, Modi N, Saugstad OD, Antetere R, NguyenBa C, Turner M, et al. Evaluating preterm care across Europe using the eNewborn European Network database. Pediatr Res. 2020;88(3):484–95. Epub 2020/01/24.10.1038/s41390-020-0769-x.
3. Jensen EA, Schmidt B. Epidemiology of bronchopulmonary dysplasia. Birth Defects Res A Clin Mol Teratol. 2014;100(3):145–57. Epub 2014/03/19.10.1002/bdra.23235.
4. Jobe AH, Bancalari E. Bronchopulmonary dysplasia. Am J Respir Crit Care Med. 2001;163(7):1723–9. Epub 2001/06/13.10.1164/ajrccm.163.7.2011060.
5. Jobe AJ. The new BPD: an arrest of lung development. Pediatr Res. 1999;46(6):641–3. Epub 1999/12/10.10.1203/00006450-199912000-00007

References

6. Dassios T, Williams EE, Hickey A, Bunce C, Greenough A. Bronchopulmonary dysplasia and postnatal growth following extremely preterm birth. Arch Dis Child Fetal Neonatal Ed. 2020; Epub 2020/12/19.10.1136/archdischild-2020-320816
7. Greenough A, Lagercrantz H. Catecholamine abnormalities in transient tachypnoea of the premature newborn. J Perinat Med. 1992;20(3):223–6. Epub 1992/01/01.10.1515/jpme.1992.20.3.223.
8. McGillick EV, Te Pas AB, van den Akker T, Keus JMH, Thio M, Hooper SB. Evaluating clinical outcomes and physiological perspectives in studies investigating respiratory support for babies born at term with or at risk of transient tachypnea: a narrative review. Front Pediatr. 2022;10:878536. Epub 2022/07/12.10.3389/fped.2022.878536.
9. Sayad E, Silva-Carmona M. Meconium aspiration. StatPearls. Treasure Island (FL) ineligible companies. Disclosure: Manuel Silva-Carmona declares no relevant financial relationships with ineligible companies. 2025.
10. Zani A, Chung WK, Deprest J, Harting MT, Jancelewicz T, Kunisaki SM, et al. Congenital diaphragmatic hernia. Nat Rev Dis Primers. 2022;8(1):37. Epub 2022/06/02.10.1038/s41572-022-00362-w.
11. Dassios T, Shareef Arattu Thodika FM, Williams E, Davenport M, Nicolaides KH, Greenough A. Ventilation-to-perfusion relationships and right-to-left shunt during neonatal intensive care in infants with congenital diaphragmatic hernia. Pediatr Res. 2022; Epub 2022/03/21.10.1038/s41390-022-02001-2
12. Lewis L, Sinha I, Kang SL, Lim J, Losty PD. Long term outcomes in CDH: Cardiopulmonary outcomes and health related quality of life. J Pediatr Surg. 2022;57(11):501–9. Epub 2022/05/05.10.1016/j.jpedsurg.2022.03.020.
13. Acun C, Nusairat L, Kadri A, Nusairat A, Yeaney N, Abu Shaweesh J, et al. Pneumothorax prevalence and mortality per gestational age in the newborn. Pediatr Pulmonol. 2021;56(8):2583–8. Epub 2021/05/19.10.1002/ppul.25454.
14. Gallacher DJ, Hart K, Kotecha S. Common respiratory conditions of the newborn. Breathe. 2016;12(1):30–42. Epub 2016/04/12.10.1183/20734735.000716.

Ventilation 3

In health, for oxygenation and gas exchange to occur, atmospheric air including oxygen needs to enter the lungs and reach the gas exchange interface, at the level of the alveoli. Fresh gas enters the lungs with every spontaneous breath and the total volume of air coming through with every breath is called the tidal volume (V_T). This volume in healthy adults is in the area of 500 ml [1] and in the newborn is calculated at approximately 4–7 ml/kilogram of body weight [2].

This weight-adjustment in neonates implies large variations of the tidal volume in different-sized infants. A normal newborn infant with a birth weight of three kilograms will have a tidal volume in the range of 15 ml, while a tiny extreme preterm with a weight of 500 grams will have a six-times smaller tidal volume of approximately only 2,5 ml. These huge differences are further augmented by the relatively fixed contribution of the breathing apparatus if such an infant is intubated and invasively ventilated, where the added volume of the apparatus will be relatively fixed and in the region of 3 ml irrespective of the weight of the infant. At best there will be 1 ml for the endotracheal tube and another 2 ml for the combined volume of the flow and carbon dioxide sensors [3]. This added volume corresponds to more than 100% of the tidal volume for a very small infant, to 20% for a term well-grown infant and would be less than 1% in a healthy adult.

Tidal Volume Compartments

The tidal volume can be divided in some compartments according to their contribution to gas exchange. The actual volume that reaches the alveoli and participates in gas exchange at the level of the alveolar membrane is the ***alveolar volume***.

A significant part of the tidal volume corresponds to areas of the respiratory system that do not participate in gas exchange such as the mouth, the pharynx, the nasal cavity, the larynx, the trachea, the main bronchi and all the bronchial subdivisions down to the alveolar zone.

In the human lung the airways are divided starting from the trachea to the right and left main bronchi and further divided for a total of approximately twenty-four times [4]. Only the last three generations of the bronchial tree contain alveoli and can thus participate in gas exchange. This part of the tidal volume which does not take part in gas exchange is called the ***anatomical dead space*** and in a healthy adult it corresponds to approximately one third of the total tidal volume of 500 ml, or 150 ml [1].

Another part of the tidal volume consists of the volume of gas that reaches the alveoli at areas of the lung which are not perfused by the pulmonary circulation, and as such also do not contribute to gas exchange. This part of the tidal volume is called the ***alveolar dead space*** and in health is practically very small or negligible as the great majority of the ventilated areas are sufficiently perfused.

The aggregate of the anatomical and the alveolar dead space is called the ***physiological dead space*** (Fig. 3.1).

In patients who are intubated and mechanically ventilated, these volumes and dead spaces are slightly different. The endotracheal tube bypasses the upper airways and delivers conditioned gas straight into the trachea, so what we calculate as the "anatomical dead space" is an aggregate of the dead space of the trachea below the endotracheal tube, the airways below the trachea plus the ***instrumental or apparatus dead space*** which is the endotracheal tube, the flow sensor and (if there is one) the carbon dioxide sensor (Fig. 3.2).

Tidal Volume Compartments

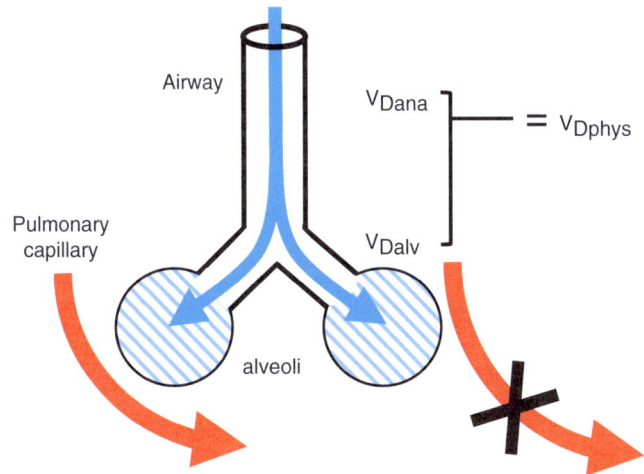

Fig. 3.1 Compartments of the tidal volume. The aggregate of the anatomical (V_{Dana}) and the alveolar (V_{Dalv}) dead space is the physiological dead space (V_{Dphys})

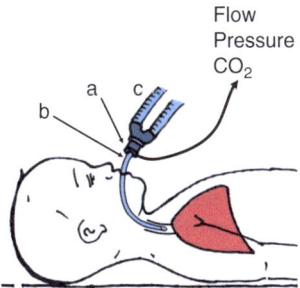

Fig. 3.2 Positions of the flow, pressure and CO_2 sensors. The sensors (a) are located between the endotracheal tube (b) and the ventilator circuit (c). The sensors, connectors and endotracheal tube all contribute to an increased apparatus dead space

The ratio of the physiological dead space to the tidal volume (V_D/V_T) is usually approximately 0.35 in healthy adults [5] and of a similar magnitude in healthy children [6] but in the newborn it can attain higher values such as 0.50 in term infants or even above 0.75 in sick premature infants with RDS or BPD on mechanical ventilation [7]. This implies that the efficiency of ventilation is overall lower in infants compared to adults and lower in sick preterm infants compared to term ones.

This ratio is higher in preterm ventilated infants compared to term ones because of the higher contribution of the anatomical dead space (both the "real" anatomical dead space and the apparatus/instrumental dead space) in smaller infants. This means that the conducting zone constitutes a larger part of the total tidal volume in the newborn and this phenomenon is even more pronounced in the very premature and small infants. Furthermore, the fixed contribution of the instrumental dead space (in the region of 3 ml as described above) is more significant in a very small infant whose total tidal volume is 2,5 ml compared to an older infant whose tidal volume is 15 or 20 ml.

We have previously used low dead space capnography to describe that the actual median value of the physiological dead space in a ventilated term infant is 3.5 ml/kg and in the preterm is 5.7 ml/kg, increasing to 6.4 ml/kg in infants with chronic disease (Fig. 3.3) [7].

The clinical significance of having a higher dead space in premature compared to term infants is that preterms will likely require a higher targeted tidal volume per kilogram when ventilated on volume targeted ventilation. For example, a targeted tidal volume of 5 ml/kg or less might be adequate for a term ventilated infant, but a preterm infant will possibly require volume targeting in the area of 6 ml/kg. These volume requirements increase further in ventilated infants with evolving or established BPD as the alveolar dead space also gradually increases reflecting a progression of the lung disease. Preterm infants with BPD who have been ventilated for long periods of time can also increase their *anatomical* dead space due to a widening of their large bronchi and the trachea (bronchomalacia, tracheomegaly). These infants might eventually need targeted tidal volumes in excess of 7 or 8 ml/kg [7].

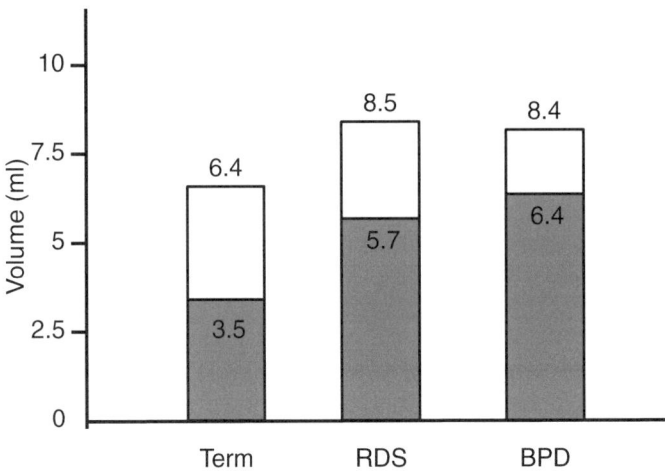

Fig. 3.3 Dead spaces in ventilated infants. The values of the physiological dead space (grey part of the column) in relation to the total tidal volume (total height of the column) are depicted in term ventilated infants, infants with respiratory distress syndrome (RDS) and evolving bronchopulmonary dysplasia (BPD) as per reference Williams [7]

If the tidal volume is multiplied with the breathing rate, the *minute ventilation* can be calculated. The part of the tidal volume which corresponds to the alveolar volume (alveolar volume times the breathing rate) is called *alveolar ventilation* and is the part that actively contributes to gas exchange. Conversely, the part of the minute ventilation which corresponds to the dead space (dead space times the breathing rate) is called *dead space ventilation* (or wasted ventilation as this part does not participate in gas exchange).

The anatomical dead space is a parameter which is usually determined by the size of the infant and the demographics. As a bedside rule of thumb, preterm ventilated infants have anatomical dead spaces in the region of 2.5 ml/kg and term born infants at 1.5 ml/kg [7]. The anatomical dead space can be larger in some infants with tracheomalacia or tracheomegaly which can be either the result of prolonged mechanical ventilation or of antenatal interventions such as the foetal endoluminal tracheal occlusion for the treatment of congenital diaphragmatic hernia [8]. These infants have typically anatomical dead spaces in the area of 3 ml/kg.

The alveolar dead space in the newborn is less influenced by anthropometric indices and is more the result of parenchymal lung disease and ongoing pulmonary injury. The actual values are much smaller, in the region of 0.5 ml/kg but the alveolar dead space is a sensitive index of disease severity, with values increasing to approximately 1 ml/kg in preterm infants with evolving BPD (Fig. 3.3) [7].

Volume Targeted Ventilation

Most ventilators of the current generation use a ***proximal flow sensor*** to measure the flow between the endotracheal tube and the ventilator circuit. The flow signal is integrated over time to present on the screen the measured volume which is delivered by the ventilator (inspiratory tidal volume—V_{Tinsp}) and the volume which is exhaled by the infant back into the expiratory limb of the ventilator circuit (expiratory tidal volume—V_{Texp}).

The ventilator uses this signal to calculate the percentage of leak as:

$$V_{Tinsp} - V_{Texp} / V_{Tinsp} \times 100$$

For example, if the set tidal volume or inspiratory tidal volume is 10 ml and the expiratory tidal volume is 8 ml, the leak will be (10 − 8 ml/10 ml) x 100 or 20%. The tidal volume reading is also used to deliver volume-targeted ventilation, as the ventilator increases or decreases the delivered pressure so that the next inflation is given by the ventilator closer to the volume target. For example, let us suppose that the ventilator is programmed to deliver a targeted volume of 10 ml. It then gives an inflation with a pressure of 20 cm H_2O and the flow sensor reads a volume of 8 ml. The next inflation will then be delivered with 10% higher pressure (22 cm H_2O). If then the read volume is 12 ml, the next inflation will be provided 10% lower so that the delivered volume is again closer to the targeted value of 10 ml (Fig. 3.4). This mechanism also explains the terminology "volume-targeted ventila-

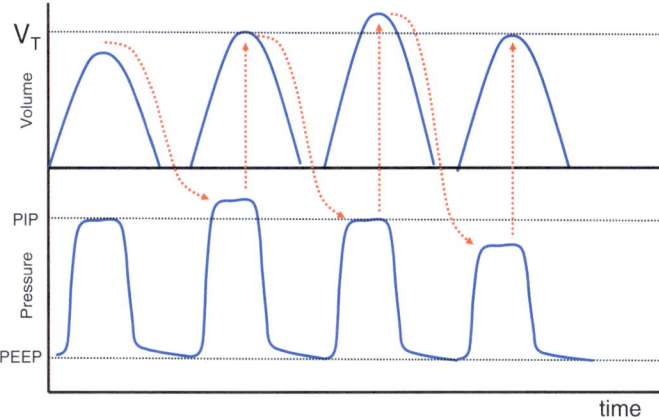

Fig. 3.4 Volume targeting. Changing the provided peak inflation pressure in response to tidal volume readings. V_T targeted tidal volume, PEEP positive end expiratory pressure, PIP peak inflation pressure

tion", as the actual tidal volume is targeted and fluctuates slightly around a set value, and is not always "guaranteed" to achieve a predefined value.

Calculation of Dead Spaces

While the tidal volume is a known parameter (and a targeted parameter in volume-targeted ventilation), the dead spaces as part of the tidal volume are more difficult to calculate at the bedside. One way to calculate these dead spaces is by combining the signals of the expired carbon dioxide with these of the expired volume in a graph which is called volumetric capnogram.

This graph (Fig. 3.5) depicts the change in the exhaled carbon dioxide versus the expired volume. In clinical practice, clinicians might be more familiar with the time capnogram which presents the exhaled carbon dioxide against time and is sometimes displayed either in intensive care monitors or on the ventilator screen. The volumetric capnogram however, is slightly different in that it is a graph of the expired CO_2 ***versus volume*** and corresponds

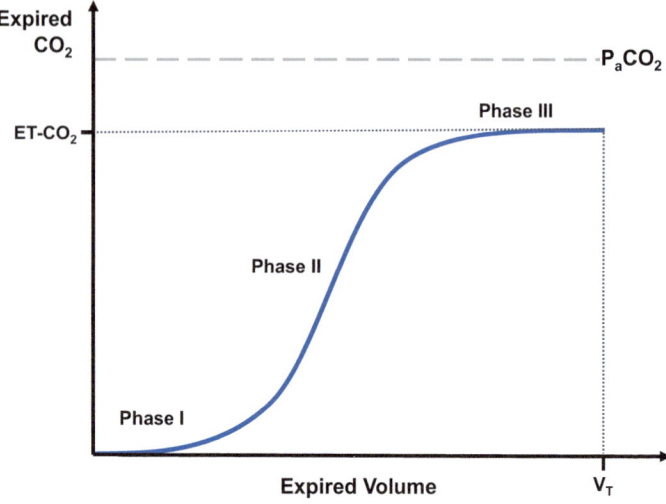

Fig. 3.5 Volumetric capnogram. The three phases of a volumetric capnogram and the points corresponding to end tidal CO_2 (ET CO_2), arterial CO_2 ($PaCO_2$) and expiratory tidal volume (V_T) are depicted

essentially to one single exhalation—or deflation if the subject is mechanically ventilated.

The volumetric capnogram consists of three phases (Fig. 3.5). The first phase which is characterised by exhalation of volume with very little or no CO_2, and corresponds to the anatomical dead space which are the non- gas exchanging parts of the lung. The second phase is characterised by a steeper rise in the expired CO_2 while the gas from airways empties out while it is being mixed with gas from the alveoli (which contain CO_2). The third phase corresponds to pure alveolar gas emptying and thus contains significant quantities of CO_2. The calculation of the dead space from the volumetric capnogram has been a historical undertaking in respiratory physiology with the first method having been described in the nineteenth century by the Danish Physiologist Christian Bohr [9]. In concept, the geometrical division of the second phase of the volumetric capnogram into two equal triangles by a vertical line to the expired volume axis, defines a volume point on the x-axis which corresponds to the anatomical dead space using the vertical projection from the point the two triangles meet the

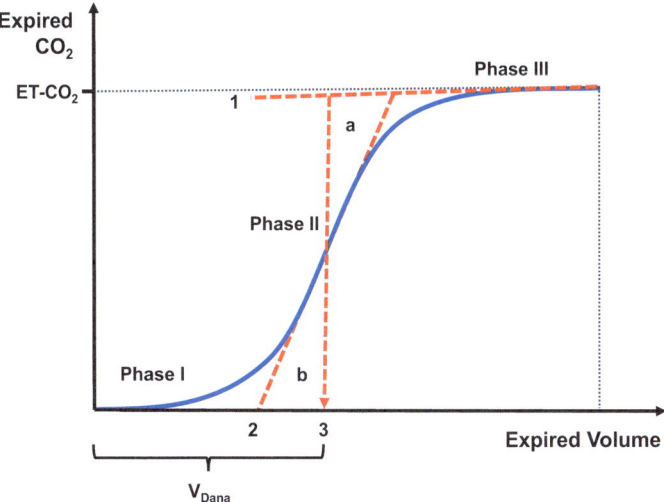

Fig. 3.6 **Volumetric capnography for the calculation of the anatomical dead space (V_{Dana}).** (1) Slope of phase III. (2) Slope of phase II. (3) A perpendicular line is projected onto the x axis so that the areas of the two triangles (a and b) are equal

carbon dioxide axis (Fig. 3.6) [10]. The concurrent knowledge of the arterial partial pressure of the CO_2 can help us calculate the total physiological dead space using a specific equation [11, 12].

It is worth remembering that the $PaCO_2$ will always be higher than the end-tidal (maximum exhaled) CO_2. This is because the alveolar CO_2 (the concentration of CO_2 in the alveoli) is thought to be very close to the arterial CO_2 provided there is no major diffusion limitation by a very abnormal and thickened alveolar membrane. During expiration, the alveolar CO_2 is diluted with the anatomical and instrumental dead space compartments which do not produce exhaled CO_2, and the actual content of the exhaled CO_2 is thus decreased to a lower value than the arterial or alveolar CO_2. The larger the dead space, the higher the dilutional effect of the dead space will be, and by knowing the difference between the arterial and the exhaled/end-tidal CO_2 one can work backwards and calculate the value of the total dead space in a specific individual.

These calculations are very delicate and prone to synchronisation error. For example, if the CO_2 and volume signals have a small time lag between them, and the CO_2 is measured one second later than the volume, this means that the volume signal will be erroneously recorded for one second while there is no exhaled carbon dioxide (and will correspond to a longer phase I in volumetric capnography), and this asynchrony will lead to an overestimation of the anatomical dead space. The opposite will happen if the volume is measured erroneously earlier compared to the carbon dioxide: it will lead to an under-estimation of the true dead space. Furthermore, in the neonatal population, where the tidal volumes are very small, there is some concern that there is re-breathing of carbon dioxide from a relatively large carbon dioxide sensor (or capnograph). As previously described, even the smallest sensor of 1 ml will constitute approximately 40% of the tidal volume in an infant of 500 g ventilated at 5 ml/kg.

Finally, the faster breathing rates in the newborn, going as high as 80–100 per minute in the smallest of the preterm ones, question whether there an alveolar plateau can always be present, which will define phase III, which in turn is required to accurately measure the end tidal CO_2 and the anatomical dead space. For all the above reasons, it has not been feasible so far for these indices to be displayed in real-time on the screen of the neonatal ventilators and the values of the dead spaces are not known to the clinicians in real time. These parameters can be calculated offline using delicate synchronisation methodology, signal aligning and quality control.

While the exact knowledge of these parameters is not currently accessible in real-life clinical neonatology, one can infer dynamic changes of these indices in the same individual by observing relative temporal changes in the difference between the arterial and the end-tidal CO_2, since these parameters are indeed readily available at the bedside. A growing difference between the arterial and end-tidal CO_2 (increasing $PaCO_2$ – $ET-CO_2$ gradient) in the same individual can be attributed to evolving parenchymal disease and an increasing alveolar dead space, as the disease progresses and the lungs are damaged in the process. The anatomical and instrumental compartments of the dead spaces are not likely to drastically change in the same ventilated subject in a brief time period, so such changes can be attributed only to evolving parenchymal disease and an increasing alveolar dead space.

Newborn infants have the unique anatomical characteristic of having a patent ductus arteriosus and foramen ovale, which in the presence of a supra-systemic pulmonary pressures can act as hydraulic decompression valves which will shunt deoxygenated blood from the high-pressure right circulation to the systemic circulation. This condition (persistent pulmonary hypertension of the newborn or PPHN) can also affect term born infants with acute pathology associated with an increased pressure in the pulmonary circulation. Recirculation of CO_2-rich blood from the pulmonary circulation into the systemic circulation will also decrease the pulmonary blood flow and eventually increase the $PaCO_2$-ET-CO_2 gradient (Fig. 3.7). In this case, this gradient does not describe evolving parenchymal lung damage and increasing

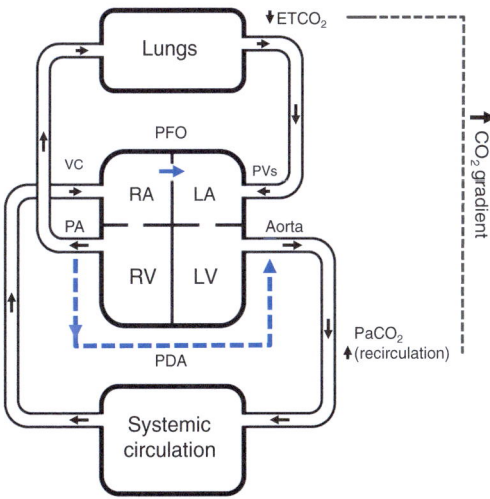

Fig. 3.7 Recirculation of CO_2 in persistent pulmonary hypertension of the newborn. Shunting increases the CO_2 content of the arterial and capillary blood and decreases pulmonary perfusion and the end-tidal CO_2. RA right atrium, RV right ventricle, LA left atrium, LV left ventricle, VC vena cavae, PA pulmonary artery, PVs pulmonary veins, PDA patent ductus arteriosus, PFO patent foramen ovale, $PaCO_2$ arterial partial pressure of CO_2, $ETCO_2$ end-tidal CO_2

alveolar dead space but can be used as an index of the magnitude of the right to left shunting. We have reported that this gradient can increase to approximately 10 mmHg in ventilated infants with persistent pulmonary hypertension and without significant lung disease, and can decrease back down to 3 mmHg after right to left shunting has ceased and PPHN has been clinically overcome [13].

Alternative Mechanisms of Gas Transport

Conventional respiratory physiology describes that gas exchange happens at the level of the alveolar membrane where the alveolar gas meets the capillary network and the gases cross the alveolar membrane by passive diffusion. Other than this *bulk gas convection*, however, some other mechanisms have also been described, particularly in relation to high-frequency ventilation, which is thought to deliver volumes in the range of the anatomical dead space or even below that dead space. These exact mechanisms are more the topic of a ventilation treatise rather than of an applied physiology book, but they include axial convection, radial diffusive mixing, coaxial flow, viscous shear, asymmetrical velocity profiles and the pendelluft effect [14]. Furthermore, some observations stemming from bench studies and real clinical data have suggested that these mechanisms might be present even at conventional respiratory rates in small preterm infants with narrow endotracheal tubes, in whom effective carbon dioxide elimination is possible with tidal volumes smaller that the dead space, possibly via spikes of fresh gas which penetrate through the dead space and create an interface of gas exchange within the conducting airways [15–17].

This is indeed a compelling theory, when however we used concurrent measurements of volumetric capnography with noninvasive measurements of ventilation to perfusion mismatch to investigate this potential phenomenon and provide insight into the location of gas exchange impairment, i.e., whether gas exchange happens solely at the alveolar membrane or in the conducting airways also, we reported that abnormal gas exchange was only associated with indices of lung disease at the alveolar level and not at the level of the airways [18].

Questions

Question 1: The physiological dead space:
  (a) Is the aggregate of the alveolar and the apparatus dead space
  (b) Can be calculated using data only collected from the flow sensor
  (c) Is the same in ml/kg in term and preterm infants
  (d) Is the volume of gas that reaches the alveoli with every breath
  (e) Can affect the value of the end tidal carbon dioxide

Question 2: The total dead space:
  (a) Is higher in ml/kg in BPD compared to RDS
  (b) Times the tidal volume equals the minute ventilation
  (c) Is not affected by the volume of the endotracheal tube and the sensor
  (d) Is increased if an infant develops tracheomegaly
  (e) Cannot change in the same infant over the course of stay

Question 3: In volume-targeted ventilation the following statements are true:
  (a) The delivered volume is measured by the flow sensor
  (b) The delivered volume is always exactly the same from inflation to inflation
  (c) The delivered volume is always exactly what we target
  (d) The delivered pressure stays the same from inflation to inflation

Question 4: For the calculation of the respiratory dead space:
  (a) We need to combine signals of pressure and flow
  (b) We can use a waveform of carbon dioxide versus time
  (c) The first phase of the capnogram corresponds to the alveolar plateau
  (d) Synchrony of the volume and carbon dioxide signals is not important
  (e) The arterial carbon dioxide is higher than the end tidal carbon dioxide

References

1. West JB. Respiratory physiology : the essentials, vol. viii. 9th ed. Philadelphia: Wolters Kluwer Health/Lippincott Williams & Wilkins; 2012. 200 pp.
2. Keszler M, Nassabeh-Montazami S, Abubakar K. Evolution of tidal volume requirement during the first 3 weeks of life in infants <800 g ventilated with volume guarantee. Arch Dis Child Fetal Neonatal Ed. 2009;94(4):F279–82. https://doi.org/10.1136/adc.2008.147157.
3. Kalous P, Kokstein Z. Instrumental dead space in neonatology, and its elimination by continuous tracheal gas insufflation during conventional ventilation. Acta Paediatr. 2003;92(5):518–24. Epub 2003/07/04.10.1080/08035250310002704.
4. Weibel ER. Morphometry of the human lung, vol. xi. Berlin: Springer; 1963. 151 pp.
5. Zimmerman MI, Miller A, Brown LK, Bhuptani A, Sloane MF, Teirstein AS. Estimated vs actual values for dead space/tidal volume ratios during incremental exercise in patients evaluated for dyspnea. Chest. 1994;106(1):131–6. Epub 1994/07/01.10.1378/chest.106.1.131.
6. Kerr AA. Dead space ventilation in normal children and children with obstructive airways diease. Thorax. 1976;31(1):63–9. Epub 1976/02/01.10.1136/thx.31.1.63.
7. Williams E, Dassios T, Dixon P, Greenough A. Physiological dead space and alveolar ventilation in ventilated infants. Pediatr Res. 2022;91(1):218–22. Epub 2021/02/20.10.1038/s41390-021-01388-8.
8. Williams EE, Dassios T, Murthy V, Greenough A. Anatomical deadspace during resuscitation of infants with congenital diaphragmatic hernia. Early Human Dev. 2020;149:105150. Epub 2020/08/11.10.1016/j.earlhumdev.2020.105150.
9. Bohr C, Hasselbalch K, Krogh A. Über einen in biologischer Beziehung wichtigen Einfluss, den die Kohlensäurespannung des Blutes auf dessen Sauerstoffbindung übt. Skand Arch Physiol. 1904;16:401–12.
10. Wenzel U, Wauer RR, Schmalisch G. Comparison of different methods for dead space measurements in ventilated newborns using CO2-volume plot. Intens Care Med. 1999;25(7):705–13. Epub 1999/09/02.10.1007/s001340050933.
11. Schmalisch G. Current methodological and technical limitations of time and volumetric capnography in newborns. Biomed Eng Online. 2016;15(1):104. https://doi.org/10.1186/s12938-016-0228-4.
12. Fowler WS. Lung function studies; the respiratory dead space. Am J Physiol. 1948;154(3):405–16. https://doi.org/10.1152/ajplegacy.1948.154.3.405.
13. Williams EE, Bednarczuk N, Nanjundappa M, Greenough A, Dassios T. Monitoring persistent pulmonary hypertension of the newborn using the arterial to end tidal carbon dioxide gradient. J Clin Monit Comput. 2024;38(2):463–7. Epub 2023/12/27.10.1007/s10877-023-01105-2.

References

14. Pillow JJ. High-frequency oscillatory ventilation: mechanisms of gas exchange and lung mechanics. Critic Care Med. 2005;33(3 Suppl):S135–41. Epub 2005/03/09.10.1097/01.ccm.0000155789. 52984.b7.
15. Keszler M, Montaner MB, Abubakar K. Effective ventilation at conventional rates with tidal volume below instrumental dead space: a bench study. Arch Dis Child Fetal Neonatal Ed. 2012;97(3):F188–92. Epub 2011/11/22.10.1136/archdischild-2011-300647.
16. Hurley EH, Keszler M. Effect of inspiratory flow rate on the efficiency of carbon dioxide removal at tidal volumes below instrumental dead space. Arch Dis Child Fetal Neonatal Ed. 2017;102(2):F126–F30. Epub 2016/08/16.10.1136/archdischild-2015-309636.
17. Nassabeh-Montazami S, Abubakar KM, Keszler M. The impact of instrumental dead-space in volume-targeted ventilation of the extremely low birth weight (ELBW) infant. Pediatr Pulmonol. 2009;44(2):128–33. Epub 2008/12/09.10.1002/ppul.20954.
18. Dassios T, Williams EE, Jones JG, Greenough A. Pathophysiology of gas exchange impairment in extreme prematurity: insights from combining volumetric capnography and measurements of ventilation/perfusion ratio. Front Pediatr. 2023;11:1094855. Epub 2023/04/04.10.3389/fped.2023.1094855.

Diffusion

After atmospheric gas has been transferred to the alveoli, the next step in the cascade of oxygen transfer from ambient air to the tissues is the diffusion across the alveolar membrane. The alveolar membrane is the interface where gas exchange occurs: the blood-gas barrier through which oxygen and carbon dioxide are transferred. There is no active transfer of gases in the human lung and the movement of both these gases takes place by passive diffusion due to the difference in their concentration across the two sides of the alveolar membrane.

The diffusion of gases at the alveolar level is described by Fick's first law of diffusion. This law states that the rate of diffusion across a membrane or a barrier is proportional to the surface area, the difference in the concentration of the gas across the two sides of the membrane and inversely proportional to the thickness of the membrane [1]:

$$V_A = \frac{S_A}{d} D (P_1 - P_2)$$

Where V_A is the alveolar ventilation, S_A is the alveolar surface area, d the thickness of the alveolar membrane, D the diffusion coefficient of the transported gas and $P_1 - P_2$ the pressure difference of the gas across the two sides of the alveolar membrane.

As described in Fick's law, diffusion is dependent on the diffusion coefficient (D or diffusivity) of the gas, a physical constant which is dependent on the physical properties of the specific gas such as the solubility and the molecular weight.

From the gases that diffuse trough the alveolar membrane, carbon dioxide has a high solubility and its diffusion rarely manifests as a significant clinical problem, compared to oxygen. Furthermore, the concentration of carbon dioxide in the atmosphere is practically negligible and the pressure gradient of CO_2 between the pulmonary blood and atmospheric air is high, which facilitates diffusion and the subsequent clearance of CO_2 from the pulmonary circulation.

In the case of oxygen, diffusion is a more complicated process. The diffusion coefficient of oxygen is lower because of a smaller solubility of the gas, and venous blood still contains some residual oxygen, which decreases the oxygen pressure difference across the two sides of the membrane.

Alveolar-Arterial Gradient

The difference in the partial pressure of oxygen across the two sides of the respiratory membrane can also be expressed or approximated as the alveolar-arterial gradient or A-a gradient, as it is a common convention in respiratory physiology to denote gases with capital letters and blood concentrations in small case letters.

In healthy individuals, the A-a gradient is very small as oxygen diffuses freely and equalises/equilibrates across the alveolar membrane. In health, the actual value would be expected to be close to zero and not exactly zero, as a small gradient is present due to some heterogeneity in perfusion and ventilation in the apical against the basal sections of the lung [2]. Practically, and since the A-a gradient increases with age, a healthy 20 year old would be expected to have a gradient of approximately 8 mmHg which will increase with age to reach a value of nearly 15 mmHg at fifty years of age [2]. In healthy term-born infants there is increased pulmonary vascular resistance and intrapulmonary shunting and this gradient has been reported to be more than three times higher

[3]. The value of the A-a oxygen gradient can exceed 60 mmHg in sick infants with RDS or chronic lung disease [4].

The A-a gradient has important clinical utility as it can help in the differential diagnosis of the aetiology of hypoxemia (Fig. 4.1). If the A-a gradient is normal, hypoxia is likely due to a low oxygen content entering the alveoli, which can originate from a reduced oxygen level in the air (as happens in high-altitude environments) or more commonly due to hypoventilation. Hypoventilation can arise from a depressed respiratory drive from the central nervous system, from severe apnoea or some degree of airway obstruction. However, if the gradient is elevated, hypoxia may be the result of ventilation to perfusion imbalance (V_A/Q mismatch). Another potential cause of hypoxaemia is the passing of deoxygenated pulmonary blood into the systemic circulation without being exposed to ventilation and oxygen (intrapulmonary shunt). Administering 100% oxygen may help differentiate between V_A/Q mismatch and intrapulmonary shunt, as oxygenation typically improves in V_A/Q mismatch but not in the case of intrapulmonary shunt [5].

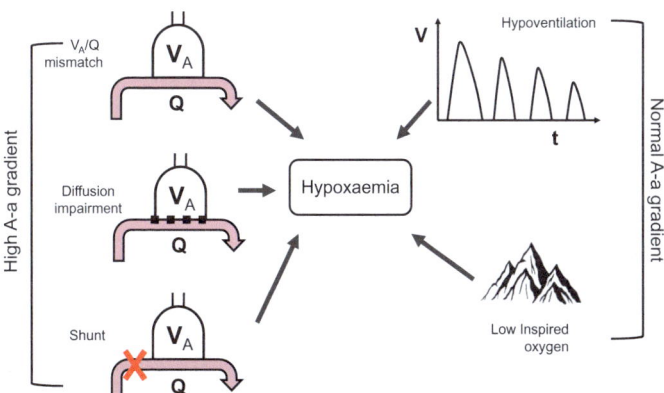

Fig. 4.1 The A-a gradient and possible aetiologies of hypoxaemia. Hypoventilation and low inspired oxygen are associated with a normal gradient, while V_A/Q mismatch, diffusion impairment or intrapulmonary shunt are associated with a high gradient

In healthy term infants the AaG attains a mean value of 17 mmHg at an age of 5 days and 32 mmHg in premature infants measured later in the first two-three months of life [6]. In disease, however, the AaG is considerably higher, going as high as approximately 200 mmHg in extremely premature ventilated infants and rising to 400 mmHg in the same infants after they develop complications such as pulmonary interstitial emphysema, a disease which by definition would impact on the diffusion capacity of the preterm lung [7]. We have also recently reported that the AaG is markedly elevated in infants with pulmonary hypertension in the context of bronchopulmonary dysplasia, highlighting a mixed origin of the encountered hypoxaemia resulting from a combination of ventilation to perfusion mismatch, right-to-left shunting and diffusion limitation [8].

Decreased Surface Area and Thickened Membrane

For both oxygen and carbon dioxide (and any other gas), diffusion can also be impaired by a decreased alveolar surface area and a thickened alveolar membrane, as described in Fick's law.

In the neonatal population, a thickened alveolar membrane has been reported in some post mortem anatomical studies of premature infants who died because of severe respiratory distress syndrome or evolving bronchopulmonary dysplasia. These studies highlighted significant interstitial disease and an increased membrane thickness in these infants compared to healthy term infants [9].

Other than a thickened barrier, the major pathophysiological process which would affect diffusion especially in the prematurely-born population is a decreased alveolar surface area [9]. Alveolarisation (the process of developing new alveoli) starts in utero and continues in postnatal life and into early childhood—or even into adolescence according to some studies [10] to reach the final number of alveoli. The aggregate of all the alveoli corresponds to a very large alveolar surface area in the range of 50 m^2 in a healthy adult. The cardinal respiratory pathophysiological event which occurs following premature birth is an arrest or a marked decrease in the process of alveolarisation, which eventually leads to fewer and larger alveolar sacs and a significantly decreased total alveolar surface area.

It is interesting to note that despite supportive postnatal extrauterine care, it seems that the alveolar surface area is significantly determined by the gestational age *at birth*, and there is limited postnatal catch up growth [11]. Using mathematical modelling and morphological data adapted from animal studies we have constructed a model to quantify this decrease in the alveolar surface area and concluded that while a healthy term newborn might have a surface area of approximately 5 m^2, a preterm born infant when measured at an equivalent postconceptional age might have a surface area of 1 m^2—which is fivefold smaller (Fig. 4.2) [12].

Another neonatal respiratory condition where significant lung underdevelopment might lead to a critically decreased alveolar surface area is congenital diaphragmatic hernia. In this condition the presence in utero of a large intestinal mass in the thorax disrupts normal lung development and leads to sometimes severe pulmonary hypoplasia secondary to a markedly decreased alveolar surface area. The condition is also characterised by significant vascular abnormalities and the ensuing imbalances in the matching of ventilation to perfusion which will be discussed in a later chapter.

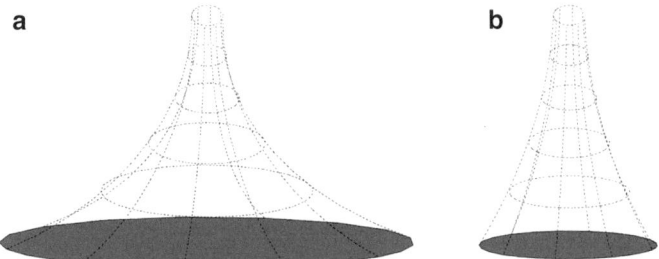

Fig. 4.2 Alveolar arrest. Schematic representation of the relative size of the alveolar surface area in term (**a**) and prematurely-born (**b**) infants

Questions

Question 1: The diffusion of a gas across a membrane:
  (a) Is inversely proportional to the surface of the membrane
  (b) Is proportional to the thickness of the membrane
  (c) Does not depend on the concentration of the gas at either side of the membrane
  (d) Is passive for oxygen and carbon dioxide across the alveolar membrane
  (e) Is independent of the properties of the gas which is being transferred

Question 2: The alveolar arterial gradient
  (a) Decreases with increasing lung disease severity
  (b) Can help in the differential diagnosis of hypoxaemia
  (c) Is nearly zero in severe bronchopulmonary dysplasia
  (d) Is increased in hypoventilation

Question 3: Oxygen diffusion in the newborn:
  (a) Is not affected by a thickened alveolar membrane
  (b) Is limited in bronchopulmonary dysplasia because of an increase in the alveolar surface area
  (c) Is not affected in congenital diaphragmatic hernia
  (d) Is normalised after discharge in preterm infants as they catch up to normal values of avleolar surface area
  (e) Is limited when the alveolar surface area is reduced

References

1. West JB. Respiratory physiology : the essentials, vol. viii. 9th ed. Philadelphia: Wolters Kluwer Health/Lippincott Williams & Wilkins; 2012. 200 pp.
2. Hantzidiamantis PJ, Amaro E. Physiology, alveolar to arterial oxygen gradient. StatPearls. Treasure Island (FL) with ineligible companies. Disclosure: Eric Amaro declares no relevant financial relationships with ineligible companies. 2024.
3. Nelson NM, Prod'hom LS, Cherry RB, Lipsitz PJ, Smith CA. Pulmonary function in the newborn infant: the alveolar-arterial oxygen gradient. J Appl Physiol. 1963;18(3):534–8. Epub 1963/05/01.10.1152/jappl.1963.18.3.534.

References

4. McCann EM, Goldman SL, Brady JP. Pulmonary function in the sick newborn infant. Pediatr Res. 1987;21(4):313–25. Epub 1987/04/01.10.1203/00006450-198704000-00001.
5. Bhutta BS, Alghoula F, Berim I. Hypoxia. StatPearls. Treasure Island (FL) ineligible companies. Disclosure: Faysal Alghoula declares no relevant financial relationships with ineligible companies. Disclosure: Ilya Berim declares no relevant financial relationships with ineligible companies. 2025.
6. Bourbon J, Boucherat O, Chailley-Heu B, Delacourt C. Control mechanisms of lung alveolar development and their disorders in bronchopulmonary dysplasia. Pediatr Res. 2005;57(5 Pt 2):38R–46R. Epub 2005/04/09.10.1203/01.PDR.0000159630.35883.BE.
7. Williams E, Dassios T, Clarke P, Chowdhury O, Greenough A. Predictors of outcome of prematurely born infants with pulmonary interstitial emphysema. Acta Paediatr. 2019;108(1):106–11. Epub 2018/05/14.10.1111/apa.14400.
8. Jeffreys E, Arattu Thodika FMS, Bell A, Greenough A, Dassios T. Mechanisms of hypoxaemia in late pulmonary hypertension associated with bronchopulmonary dysplasia. J Perinatal Med. 2025;53(4):549–51. Epub 2025/04/01.10.1515/jpm-2025-0058.
9. Margraf LR, Tomashefski JF Jr, Bruce MC, Dahms BB. Morphometric analysis of the lung in bronchopulmonary dysplasia. Am Rev Respir Dis. 1991;143(2):391–400. Epub 1991/02/11.10.1164/ajrccm/143.2.391.
10. Narayanan M, Owers-Bradley J, Beardsmore CS, Mada M, Ball I, Garipov R, et al. Alveolarization continues during childhood and adolescence: new evidence from helium-3 magnetic resonance. Am J Respir Critic Care Med. 2012;185(2):186–91. Epub 2011/11/11.10.1164/rccm.201107-1348OC
11. Sobonya RE, Logvinoff MM, Taussig LM, Theriault A. Morphometric analysis of the lung in prolonged bronchopulmonary dysplasia. Pediatr Res. 1982;16(11):969–72. https://doi.org/10.1203/00006450-198211000-00014.
12. Dassios T, Dassios KG, Dassios G. Functional morphometry for the estimation of the alveolar surface area in prematurely-born infants. Respir Physiol Neurobiol. 2018;254:49–54. https://doi.org/10.1016/j.resp.2018.04.008.

Perfusion

5

Following the diffusion of gases across the alveolar membrane, the next step in the cascade of oxygen transfer is the transport of the oxygen-rich blood via the pulmonary circulation to the left circulation and from there to the peripheral tissues.

Pulmonary blood flow can be measured using *Fick's principle* which requires a relatively invasive methodology as it requires knowledge of the oxygen content in the pulmonary artery and thus entails arterial catheterisation. Alternatively, it can be measured using Doppler echocardiography by integrating the flow velocity in the pulmonary artery versus time [1]. Using such methodology we have reported a median pulmonary perfusion of 350 ml/min in extremely preterm infants measured in the first week of life [2]. The pulmonary circulation, compared to the systemic, operates at lower pressures and against vessels with a lower vascular resistance. This is logical, considering that the pulmonary circulation has a shorter path to travel, less vascular resistance to overcome and needs to be in a relative pressure equilibrium with the low-pressure alveolar compartment for efficient gas exchange to occur.

Pulmonary Vascular Resistance

Beyond the neonatal period and in healthy conditions, the pulmonary circulation operates in series with the systemic circulation and oxygenated blood is delivered from the lungs to the left ventricle and from there to the tissues without mixing of blood between the two circulations. In the neonatal period, however, there is an anatomical setup for potential communication between the two circulations via the ductus arteriosus and the foramen ovale. This has major pathophysiological implications, because an increase of the pulmonary vascular resistance and/or pulmonary artery pressure would automatically cause shunting of deoxygenated blood from the right into the left circulation and the delivery to the tissues of a mixture of oxygenated and deoxygenated blood which will clinically manifest as hypoxaemia and cyanosis. We have already described this condition as persistent pulmonary hypertension of the newborn (PPHN) and because of this potential for direct cardiac right-to-left shunting, it differs significantly compared to adult pulmonary hypertension. In adults pulmonary hypertension will manifest with structural changes in the right heart and right ventricular failure which will lead to a chronic disease of the right ventricle broadly described by the term *cor pulmonale*.

Clinically, the major parameter of pathophysiological importance for neonatal pulmonary hypertension is the pulmonary vascular resistance (PVR). The vascular resistance is defined by the ratio of the pressure difference at the two ends of a vessel, divided by the flow through this vessel. It follows that a high pressure gradient between the pulmonary artery and the pulmonary veins would cause a higher resistance state and this is why the pressure difference between these two major vessels is under normal conditions considerably smaller compared to the systemic circulation (Fig. 5.1).

In utero, the PVR remains high as the intrauterine environment of the foetus is hypoxic compared to the extrauterine one. If it was physically possible to attach a probe and measure the transcutaneous oxygen saturation of a foetus in the womb, we would get val-

Pulmonary Vascular Resistance

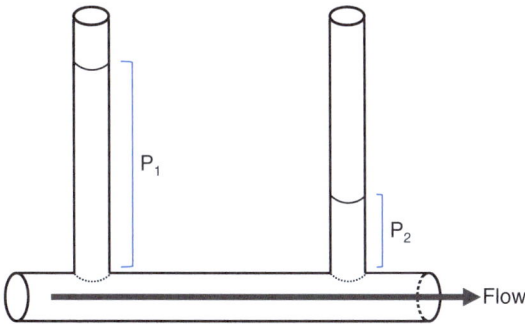

Fig. 5.1 **Estimation of vascular resistance**. The vascular resistance can be calculated as the difference in the pressures (P_1–P_2) divided by the flow across the tube

ues in the range of 60% [3] which are significantly lower compared to the normal postnatal saturations of more than 90% which are achieved within a few minutes after birth [4].

The physiological and teleological objective of having a high PVR in utero is that high PVR drives the shunting of blood from the "dormant" neonatal lungs (which have not taken over from the placenta yet as an organ of oxygenation) to the left heart via the ductus arteriosus and the foramen ovale. This way the oxygenated blood from the placenta and the umbilical vein is bypassing the right circulation and via the left circulation ends up oxygenating the foetal periphery.

When the infant is born and starts breathing spontaneously using their lungs, there is a sudden increase in the concentration of oxygen that reaches the lungs causing the PVR to decrease. Other than oxygen, the factors that contribute to the decrease in the pulmonary vascular resistance at birth include the expansion of the lungs with gas, the increase in pH, and vasoactive substances such as bradykinin, some prostaglandins such as PGE1, PGA1, PGI2 (prostacyclin), PGD2 and endothelium-derived relaxing factor [5]. The pulmonary vascular bed is particularly sensitive to several "irritants" such as aspirated meconium, pneumonitis, inflammatory mediators during neonatal respiratory distress syndrome as well as numerous other biochemical and haematological abnormalities such as polycythaemia or hypercapnic acidosis. One such very potent vasoconstricting condition is hypoxia which, either isolated or in the context of co-existing lung disease, can cause persistent and severe vasoconstriction of the pulmonary circulation, an increase of the pressure in the pulmonary artery and shunting of deoxygenated blood into the systemic circulation.

Contrarily, alkalosis has a relaxing effect on the pulmonary vessels and lowers the PVR, which is the pathophysiological rationale for the historical practice of running an intentionally high pH in PPHN with bicarbonate infusion (alkalinisation) so that the PVR decreases and the PPHN shunt is reversed. This practice has long been abandoned as clinically harmful, as unfortunately significant alkalosis also dilates the cerebral circulation and is associated with intracerebral haemorrhagic complications [6].

Another clinical factor that could influence the PVR is the total lung volume during invasive ventilation. The relationship of PVR with the total lung volume follows a "U shape", with increased PVR in very low lung volumes possibly as a result of vessel collapse which accompany parenchymal atelectasis, and an equally abnormally high PVR in very high lung volumes possibly as a result of compression of the capillaries at very high distending pressures (Fig. 5.2). In between these two extremes, lies an optimal zone of ventilation where alveolar volume is optimised while the PVR is also kept at a minimum facilitating optimal lung perfusion and thus oxygenation [7].

It is possible that hypoxic vascular pulmonary vasoconstriction is the abnormal over-activation of a physiological mechanism which normally serves (as described above) the diversion of blood in utero away from the foetal pulmonary circulation. Hypoxic vascular vasoconstriction might also serve as a protective mechanism in inhomogeneous lung disease aiming to mitigate hypoxia by diverting blood flow away from the hypoxic lung units while well-oxygenated parts of the lung retain their perfusion and thus overall ventilation to perfusion matching is preserved and oxygenation is protected.

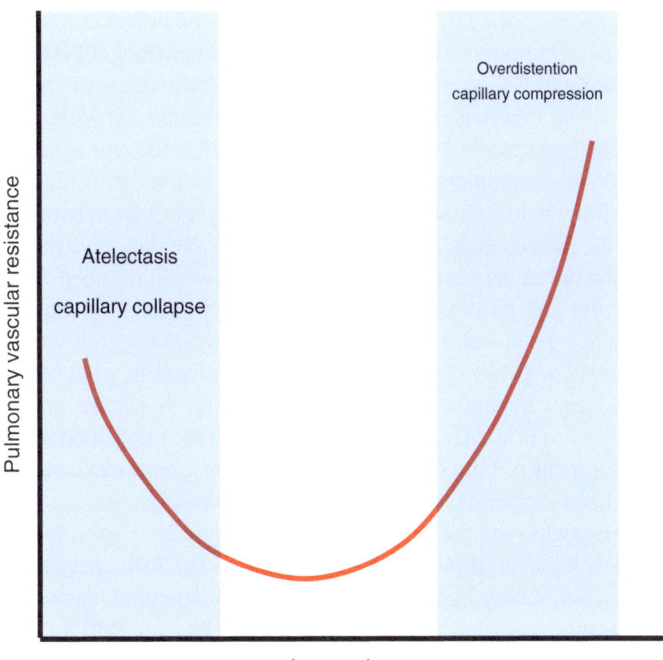

Fig. 5.2 Effect of lung volume on pulmonary vascular resistance. Low lung volumes in the range of atelectasis are associated with increased vascular resistance via alveolar collapse. High lung volumes in the range of overdistention are associated with increased pulmonary vascular resistance via vascular compression

"Fixed" Pulmonary Hypertension: Congenital Diaphragmatic Hernia—Bronchopulmonary Dysplasia

Pulmonary hypertension can occur acutely due to numerous acute conditions or sometimes without an identified cause (idiopathic pulmonary hypertension) but can also be the result of *chronic* hypoxic diseases such as bronchopulmonary dysplasia or congenital diaphragmatic hernia (CDH).

In the case of CDH, pulmonary hypertension differs compared to persistent pulmonary hypertension of the newborn (PPHN) in that while PPHN is an acute increase (or failure to decrease) in the PVR of an otherwise normal pulmonary vascular network, pulmonary hypertension in the context of CDH is the end result of a prolonged intrauterine pulmonary hypoplasia which is characterised by maldevelopment of the pulmonary vascular network in both the macroscopic and microscopic levels. Studies have shown that the pulmonary arterioles in CDH are fewer in number, more muscular and have a thicker wall compared to healthy controls [8]. Since the alveolar and vascular parts of the lungs grow simultaneously, it is not completely clear whether and to what extent pulmonary hypoplasia induces pulmonary hypertension or whether it is primarily vascular factors which are implicated in the development of lung hypoplasia. Interestingly, both mechanisms have been supported by experimental evidence [8].

Irrespective of the pathophysiological process underpinning the development of pulmonary hypertension in CDH, once established, pulmonary hypertension is of a fixed/structural nature and is thus more resistant to treatment compared to acute PPHN [9]. It is also useful to remember that since the systemic and the pulmonary circulations are meant to be working in series postnatally, an underdeveloped pulmonary vascular network in CDH which consists of fewer and "stiffer" pulmonary vessels will operate as a "pressure bottleneck" when receiving the less affected systemic supply and, even simply from the hydraulics perspective, will generate a higher pressure compared to the systemic circulation.

CDH is one of the best studied neonatal diseases for pulmonary hypertension and can serve as a model to offer valuable insights into the pathophysiology and progression of pulmonary hypertension across different pathologies. Less studied and heterogeneous disorders which are also characterised by pulmonary hypoplasia such as prolonged rupture of membranes, oligohydramnios or renal agenesis are also accompanied by varying degrees of pulmonary hypertension which are the result of a similar intrauterine process [10]. The time frame of the intrauterine pathology however, might also play a part, as severe anhydramnios and renal agenesis might be associated with more severe and refractory pulmonary hypertension compared to milder cases of oligohydramnios secondary to prolonged rupture of membranes which occurred later in the course of gestation [11].

Similarly to the intrauterine development of pulmonary hypertension in CDH, severe structural pulmonary hypertension can also occur *postnatally* in prematurely born infants, especially in the ones who fall into the chronic phase of preterm neonatal respiratory disease - captured by the term bronchopulmonary dysplasia (BPD). Approximately one third of infants with BPD will develop some degree of pulmonary hypertension [12]. BPD is characterised by arrested or severely retarded postnatal alveolarisation and lung growth [13]. This eventually leads to a simplified lung structure, with fewer and larger alveoli which correspond to a drastically smaller alveolar surface area, but also to a smaller pulmonary vascular network [13]. PH in BPD is characterised by a decreased vascular surface area and increased PVR which can lead to right ventricular dysfunction and hypertrophy [12].

The development of pulmonary hypertension in the context of BPD can attain severe and persistent forms and can be associated with higher mortality and overall morbidity. For example, while preterm infants with BPD without pulmonary hypertension had a mortality of less than 2%, development of pulmonary hypertension after 28 days of life was associated with an increased mortality up to 36% [14]. Similarly to CDH, this form of pulmonary hypertension is fixed and structural in nature and does not swiftly improve with aggressive cardiorespiratory management but can be refractory and often persists for months after discharge from neonatal care [15].

Pulmonary hypertension in BPD can also be complicated by further structural developmental anomalies such as pulmonary vein stenosis, a condition which arises from abnormal development and remodelling of the pulmonary vessels mediated by inflammatory factors, which can lead to right heart failure and pericardial effusion with impaired fluid reabsorption with a very poor prognosis and high mortality [16].

The presence and contribution of open communications between the right and left circulation (patent ductus arteriosus and foramen ovale) in the context of severe pulmonary hypertension is a matter of consideration, as theoretically an open and haemodynamically significant ductus arteriosus would worsen pulmonary hypertension, because an open ductus tends to equalise pressures between the systemic and pulmonary circulation, which means that the pressures in the pulmonary circulation remain high and this contributes to high PVR and helps the development of pulmonary hypertension [17]. In the long term, increased flow and higher pressures in the pulmonary circulation via an open ductus are associated with morphological vascular changes in the pulmonary vessels such as intimal thickening, fibrosis and worsened right to left shunting [18].

Intrapulmonary Shunting

Other than cardiac right-to-left shunting, it is possible that deoxygenated blood can bypass normal lung oxygenation through the perfusion of pulmonary segments which are not ventilated and thus not exposed to oxygen. This means that the blood in these areas will remain deoxygenated and will return to the systemic circulation without having been enriched with oxygen. This is called ***intrapulmonary shunting***. A small degree (approximately 5%) of intrapulmonary shunting is normal in the newborn [19] and is thought to occur secondary to immature smooth muscle and cartilage development in the bronchial wall of newborn infants which might cause a small proportion of the small airways of the newborn infants to collapse [20] or might also be the result of functional arterio-pulmonary collateral vessels between the bronchial and pulmonary arteries [21].

The magnitude of intrapulmonary shunting in *diseased* neonatal lungs, however, is considerably higher with a median shunt in the range of 12% in premature infants with BPD [22], 19% in infants with pulmonary interstitial emphysema [23], or 10% in infants with congenital diaphragmatic hernia [24]. Some interventions can decrease intrapulmonary shunting in the clinical setting such as the administration of intravenous caffeine which has been associated with a halving of the median shunt from 8% to 4% in

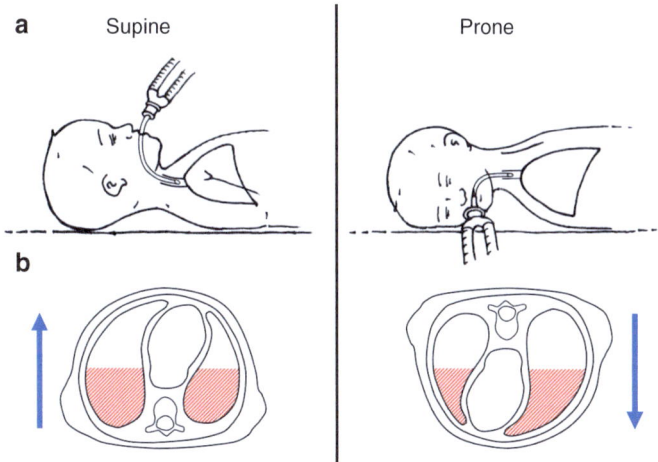

Fig. 5.3 Ventilation differences in the prone and supine positions. The dependent areas are marked with diagonal lines. The non-dependent areas are larger in prone compared to supine position

ventilated preterm infants possibly via the diuretic action of caffeine which might offload interstitial fluid thereby decreasing shunting [25]. Prone positioning has also been associated with a significant but modest decrease in right-to-left shunting in ventilated infants possibly via an improved ventilation of the previously atelectatic large dorsal lung segments which open up to adequate ventilation while remaining sufficiently perfused, decreasing thus the degree of intrapulmonary shunting (Fig. 5.3) [26].

Ductus Arteriosus

Although not primarily a respiratory disease, a large patent ductus arteriosus (haemodynamically significant as it is called to mark it as clinically important), can also affect pulmonary perfusion and respiratory function. If there is significant left to right shunting of blood from an open ductus arteriosus this might lead to increased pulmonary perfusion and subsequent pulmonary oedema which will impact respiratory status and increase the need for mechanical ventilation [27]. This condition will manifest clinically as a respiratory deterioration, inability to wean off invasive respiratory support and radiographically as pulmonary congestion (what clinicians would refer to as "wet lungs") and cardiomegaly as a sign of heart failure. Treatment thus of PDA with diuretic agents has the potential to offload this interstitial pulmonary fluid, improve pulmonary mechanics and improve gas exchange.

Vascular Origins of Respiratory Disease

The traditional narrative suggests that vascular abnormalities in BPD are the result of alveolar arrest, which is the primary causative event. An intriguing theory however, has suggested mechanistic links between foetal growth retardation and BPD implying that vascular events are the main drivers of lung developmental retardation in these infants. This theory suggests that since it is well known that foetal hypoxia disrupts angiogenesis and is associated with foetal growth restriction, a similar effect of hypoxia on the developing pulmonary vasculature could also be a critical pathogenetic factor for BPD [28].

Questions

Question 1: The pulmonary circulation in the newborn:
 (a) In health operates at a higher pressure than the systemic circulation
 (b) Operates with a higher vascular resistance than the systemic circulation
 (c) Can communicate with the systemic circulation via the foramen ovale and the ductus arteriosus
 (d) When there is pulmonary hypertension, hypoxaemia will only be subclinical

Question 2: Pulmonary hypertension in bronchopulmonary dysplasia:
(a) Responds swiftly to acute management and resolves
(b) Can be accompanied by structural changes of the right ventricle
(c) Is universal in all infants with BPD
(d) Does not translate to worse clinical outcomes

Question 3: The following is true for intrapulmonary shunting:
(a) Refers to ventilated alveoli who are not perfused
(b) In health there is no shunting at all
(c) Will be increased by administration of diuretics
(d) Can be increased when there is significant parenchymal disease

References

1. Cloez JL, Schmidt KG, Birk E, Silverman NH. Determination of pulmonary to systemic blood flow ratio in children by a simplified Doppler echocardiographic method. J Am College Cardiol. 1988;11(4):825–30. Epub 1988/04/01.10.1016/0735-1097(88)90218-5.
2. Williams EE, Gareth Jones J, McCurnin D, Rudiger M, Nanjundappa M, Greenough A, et al. Functional morphometry: non-invasive estimation of the alveolar surface area in extremely preterm infants. Pediatr Res. 2023; Epub 2023/04/13.10.1038/s41390-023-02597-z.
3. Dildy GA, van den Berg PP, Katz M, Clark SL, Jongsma HW, Nijhuis JG, et al. Intrapartum fetal pulse oximetry: fetal oxygen saturation trends during labor and relation to delivery outcome. American journal of obstetrics and gynecology. 1994;171(3):679–84. Epub 1994/09/01.10.1016/0002-9378(94)90081-7
4. Kamlin CO, O'Donnell CP, Davis PG, Morley CJ. Oxygen saturation in healthy infants immediately after birth. J Pediatr. 2006;148(5):585–9. Epub 2006/06/02.10.1016/j.jpeds.2005.12.050.
5. Ghanayem NS, Gordon JB. Modulation of pulmonary vasomotor tone in the fetus and neonate. Respir Res. 2001;2(3):139–44. Epub 2001/11/01.10.1186/rr50
6. Marron MJ, Crisafi MA, Driscoll JM Jr, Wung JT, Driscoll YT, Fay TH, et al. Hearing and neurodevelopmental outcome in survivors of persistent pulmonary hypertension of the newborn. Pediatrics. 1992;90(3):392–6. Epub 1992/09/11

7. West JB. Respiratory physiology : the essentials, vol. viii. 9th ed. Philadelphia: Wolters Kluwer Health/Lippincott Williams & Wilkins; 2012. 200 pp.
8. Mohseni-Bod H, Bohn D. Pulmonary hypertension in congenital diaphragmatic hernia. Semin Pediatr Surg. 2007;16(2):126–33. Epub 2007/04/28.10.1053/j.sempedsurg.2007.01.008.
9. Pierro M, Thebaud B. Understanding and treating pulmonary hypertension in congenital diaphragmatic hernia. Semin Fetal Neonatal Med. 2014;19(6):357–63. Epub 2014/12/03.10.1016/j.siny.2014.09.008.
10. Wu CS, Chen CM, Chou HC. Pulmonary hypoplasia induced by oligohydramnios: findings from animal models and a population-based study. Pediatr Neonatol. 2017;58(1):3–7. Epub 2016/06/22.10.1016/j.pedneo.2016.04.001.
11. de Waal K, Kluckow M. Prolonged rupture of membranes and pulmonary hypoplasia in very preterm infants: pathophysiology and guided treatment. J Pediatr. 2015;166(5):1113–20. Epub 2015/02/15.10.1016/j.jpeds.2015.01.015.
12. Berkelhamer SK, Mestan KK, Steinhorn RH. Pulmonary hypertension in bronchopulmonary dysplasia. Semin Perinatol. 2013;37(2):124–31. Epub 2013/04/16.10.1053/j.semperi.2013.01.009.
13. Coalson JJ. Pathology of new bronchopulmonary dysplasia. Semin Neonatol. 2003;8(1):73–81. https://doi.org/10.1016/s1084-2756(02)00193-8.
14. Arattu Thodika FMS, Nanjundappa M, Dassios T, Bell A, Greenough A. Pulmonary hypertension in infants with bronchopulmonary dysplasia: risk factors, mortality and duration of hospitalisation. J Perinatal Med. 2022;50(3):327–33. Epub 2021/12/01.10.1515/jpm-2021-0366.
15. Altit G, Bhombal S, Hopper RK, Tacy TA, Feinstein J. Death or resolution: the "natural history" of pulmonary hypertension in bronchopulmonary dysplasia. J Perinatol. 2019;39(3):415–25. Epub 2019/01/09.10.1038/s41372-018-0303-8.
16. Williams EE, Nanjundappa M, Babla K, Wong J, Dassios T, Greenough A. Pericardial effusion and pulmonary vein stenosis in severe bronchopulmonary dysplasia. Arch Dis Child Fetal Neonatal Ed. 2022;107(4):447. Epub 2021/04/15.10.1136/archdischild-2021-321830.
17. Chinawa JM, Chukwu BF, Chinawa AT, Duru CO. The effects of ductal size on the severity of pulmonary hypertension in children with patent ductus arteriosus (PDA): a multi-center study. BMC Pulmonary Med. 2021;21(1):79. Epub 2021/03/06.10.1186/s12890-021-01449-y.
18. Philip R, Lamba V, Talati A, Sathanandam S. Pulmonary hypertension with prolonged patency of the ductus arteriosus in preterm infants. Children. 2020;7(9) Epub 2020/09/20.10.3390/children7090139.
19. Dassios T, Ali K, Rossor T, Greenough A. Using the fetal oxyhaemoglobin dissociation curve to calculate the ventilation/perfusion ratio and right to left shunt in healthy newborn infants. J Clin Monit Comput. 2019;33(3):545–6. https://doi.org/10.1007/s10877-018-0168-6.

20. Sinclair-Smith CC, Emery JL, Gadsdon D, Dinsdale F, Baddeley J. Cartilage in children's lungs: a quantitative assessment using the right middle lobe. Thorax. 1976;31(1):40–3. Epub 1976/02/01.10.1136/thx.31.1.40.
21. Poets CF, Samuels MP, Southall DP. Potential role of intrapulmonary shunting in the genesis of hypoxemic episodes in infants and young children. Pediatrics. 1992;90(3):385–91. Epub 1992/09/01
22. Svedenkrans J, Stoecklin B, Jones JG, Doherty DA, Pillow JJ. Physiology and predictors of impaired gas exchange in infants with bronchopulmonary dysplasia. Am J Respir Critic Care Med. 2019; https://doi.org/10.1164/rccm.201810-2037OC.
23. Williams E, Dassios T, Clarke P, Chowdhury O, Greenough A. Predictors of outcome of prematurely born infants with pulmonary interstitial emphysema. Acta Paediatr. 2019;108(1):106–11. Epub 2018/05/14.10.1111/apa.14400.
24. Dassios T, Shareef Arattu Thodika FM, Williams E, Davenport M, Nicolaides KH, Greenough A. Ventilation-to-perfusion relationships and right-to-left shunt during neonatal intensive care in infants with congenital diaphragmatic hernia. Pediatr Res. 2022; Epub 2022/03/21.10.1038/s41390-022-02001-2.
25. Kaltsogianni O, Bhat R, Greenough A, Dassios T. Temporal effects of caffeine on intrapulmonary shunt in preterm ventilated infants. J Perinatal Med. 2024;52(5):556–60. Epub 2024/03/15.10.1515/jpm-2023-0492
26. Barka K, Papachatzi E, Fouzas S, Dimitriou G, Dassios T. Respiratory function in ventilated newborn infants nursed prone and supine. Pediatr Pulmonol. 2025;60(4):e71075. Epub 2025/04/01.10.1002/ppul.71075
27. Clyman RI, Hills NK. Patent ductus arteriosus (PDA) and pulmonary morbidity: can early targeted pharmacologic PDA treatment decrease the risk of bronchopulmonary dysplasia? Semin Perinatol. 2023;47(2):151718. Epub 2023/03/08.10.1016/j.semperi.2023.151718.
28. Sehgal A, Dassios T, Nold MF, Nold-Petry CA, Greenough A. Fetal growth restriction and neonatal-pediatric lung diseases: vascular mechanistic links and therapeutic directions. Paediatr Respir Rev. 2022;44:19–30. Epub 2022/12/13.10.1016/j.prrv.2022.09.002.

Ventilation to Perfusion relationships

6

In health, alveolar ventilation and pulmonary perfusion are balanced so that oxygen is delivered to the peripheral tissues at an optimal and stable rate. The gases are transferred via the airways down to the alveolar membrane where they meet the pulmonary capillaries and gas exchange occurs. These two sides of the respiratory system under healthy conditions operate with matching rates: the rate of delivery of the volume of gas to the alveoli is equal to the rate by which the pulmonary vessels supply the blood-gas barrier with blood. This balanced condition is called ventilation to perfusion (V_A/Q) matching and if one was to quantify this matching via a ratio of the ventilation to perfusion, this ratio in health would be close to 1, and usually above 0.8 [1]. For example, a newborn infant breathing at a rate of 50 breaths per minute and with a tidal volume of 20 ml will have a minute ventilation of 1000 ml. The amount of blood reaching his lungs would be his heart rate (120 bpm) times the stroke volume (10 ml) if his right and left circulations operate in series, or 1200 ml. The V_A/Q would then be 1000 ml/1200 ml or 0.83, which is a normal value.

© The Author(s), under exclusive license to Springer Nature Switzerland AG 2025
T. Dassios, *Clinical Respiratory Physiology of the Newborn*, In Clinical Practice, https://doi.org/10.1007/978-3-032-05738-9_6

Ventilation to Perfusion Mismatch

If the balance between ventilation and perfusion is disrupted for any reason relating to either abnormally high or low alveolar ventilation or pulmonary perfusion, the matching will be affected and the ratio will attain abnormal values, either higher or lower. This condition is called ventilation to perfusion mismatch and has major implications for the delivery of oxygen to the tissues as it is the commonest and most important cause of hypoxaemia. We briefly discussed in the chapter of diffusion that hypoxaemia can be attributed to a number of mechanisms such as hypoventilation, environmental hypoxia with a low inspired oxygen concentration such as in high altitude, right-to-left shunting, diffusion limitation or ventilation to perfusion mismatch.

V_A/Q mismatch can occur via a number of mechanisms:

- If alveolar ventilation is impaired more severely than perfusion, for example during acute parenchymal disease such as pneumonitis or respiratory distress syndrome, then the V_A part of the ratio will be more significantly decreased while the perfusion will remain relatively normal. This will give rise to a low V_A/Q ratio. The same would happen if a mucous plug or a blood clot blocked part of the airways and the lungs: the ventilation would be affected but the perfusion would remain relatively unchanged. Any pathological process which impacts ventilation by partially obstructing, narrowing or blocking the larger or the smaller airways (bronchopulmonary dysplasia, bronchiolitis, asthma etc) will also have the same effect, causing a low V_A/Q, as ventilation is affected but perfusion is preserved (Fig. 6.1) [2].
- If ventilation is not affected but perfusion is decreased then this will cause V_A/Q mismatch with a high V_A/Q ratio. A typical example of hypoxaemia secondary to a high V_A/Q is pulmonary emphysema where the destruction of the pulmonary vessels leads to a significant reduction in perfusion while ventilation remains normal or sometimes increased. In adult

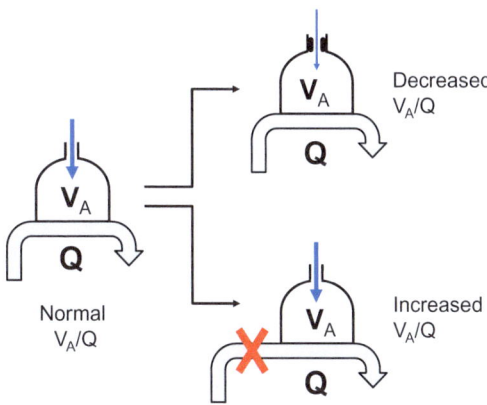

Fig. 6.1 Ventilation perfusions relationships. V_A alveolar ventilation, Q perfusion. Airway obstruction and decreased ventilation causes a decreased V_A/Q ratio (upper right diagram). Impaired perfusion and normal ventilation cause an increased V_A/Q ratio (lower right diagram)

medicine a very common cause and typical example of a high V_A/Q is pulmonary embolism which will acutely obstruct the pulmonary circulation while ventilation will not be acutely affected to the same extent. In the neonatal population, right-to-left cardiac shunting of blood from the pulmonary circulation to the systemic circulation can typically occur in persistent pulmonary circulation of the newborn. This recirculation of blood away from the lungs into the systemic circulation will acutely decrease the amount of blood received by the lungs, and will give rise to an "oligaemic" radiographic appearance of the lungs, which means that the pulmonary vessels are not delineated at all in a chest radiograph and the lungs appear completely black, while in health the large pulmonary vessels would be faintly delineated and visible.

As mentioned above, any disruption in the V_A/Q balance, either from inadequate ventilation or decreased perfusion will eventually lead to hypoxaemia.

Measuring the Ventilation to Perfusion Ratio

Measurement of the V_A/Q ratio is a cumbersome and complex process which involves nuclear medicine (V_A/Q scan) and consists of inhaling a radioactive gas to measure ventilation and injecting a radioactive tracer into the bloodstream to visualise blood flow. This investigation has rarely been used in sick newborns, or even in well infants in order to establish normal values. A large longitudinal study of surviving children with congenital diaphragmatic hernia using V_A/Q scans reported abnormally high V_A/Q ratios over time reflecting primarily abnormally low perfusion secondary to pulmonary hypertension. Interestingly this ratio increased further (became more abnormal) over the first fifteen years of life as ventilation recovered better compared to perfusion which remained abnormally low and was the main driver of the persistently high V_A/Q over time [3].

Other than nuclear medicine it is also possible to quantify V_A/Q abnormalities using non-ionising technology by newer functional magnetic resonance imaging (MRI) studies. Hyperpolarised gases such as Helium ^{3}He and Xenon ^{129}Xe can be delivered in the human lungs and MR image acquisition can be performed during a short breath-hold (lasting less than 20 s) after inhalation. The sequential images will capture the tracer gas initially in the lungs before crossing the gas-blood barrier to depict ventilation defects as the gas will not be delivered in areas which are not ventilated. The tracer gas will then cross into the pulmonary capillaries and it can be visualised in a consequent perfusion MR image which can then be combined with the ventilation image, and give information on regional ventilation to perfusion imbalances of the specific individual in the same image [4]. This methodology is only starting to be implemented in the neonatal population with some preliminary results exhibiting marked ventilation inhomogeneity and regional lung function deficits in infants with BPD [5].

These developments are evidently fascinating but realistically not accessible to the majority of the neonatal clinicians caring for sick newborns with severe lung disease, while the facilities are expensive and require significant specialised expertise. An alternative method to estimate V_A/Q relationships is the method of the oxyhaemoglobin dissociation curve (ODC), using non-invasive measurements of the provided fraction of inspired oxygen (FiO_2) and the achieved transcutaneous oxygen saturations (SpO_2) (Fig. 6.2). By using paired measurements of FiO_2 and SpO_2 the ODC of a specific individual can be constructed and with some mathematical modelling, one can work backwards using how much an individual subject's curve is misplaced relative to a normal / ideal ODC and calculate the shift of the curve to the right and the corresponding V_A/Q ratio [6, 7]. In concept, this method is based on the principle that hypoxaemia which is due to low V_A/Q will correct by increasing the provided FiO_2 [6, 7]. For example, if a ventilated infant requires 30% oxygen to maintain stable saturations at 92%, then we can increase his saturation to 96% by administering 50% oxygen. This practically means that this ODC is shifted to the right, and by calculating the degree of the right shift we can translate this to a precise V_A/Q value. The same method can also be applied to estimate the right-to-left shunting as was reported in the chapter on perfusion.

Using this method we have confirmed that, similarly to the adult population, the mean V_A/Q in healthy term infants without

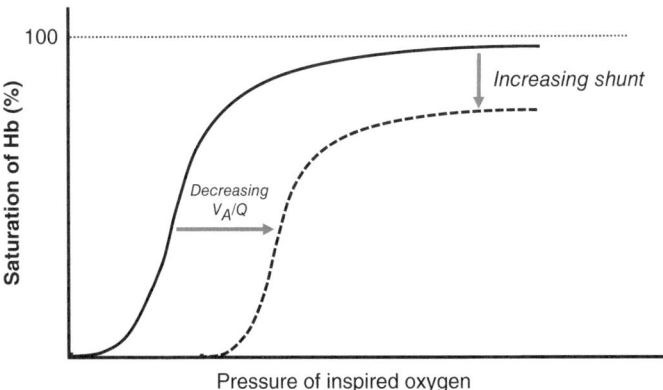

Fig. 6.2 The oxygen haemoglobin dissociation curve. The proportion of haemoglobin saturation versus the pressure of inspired oxygen. Increasing shunt displaces the curve downwards and decreasing the ventilation to perfusion ratio (V_A/Q) shifts the curve to the right

respiratory disease was 0.84 with a standard deviation of 0.18 [8]. We have reported a median value of V_A/Q of 0.40 in sick newborns with evolving BPD [9] and similar values have also been independently reported in other large independent cohorts where V_A/Q inequality tended to worsen with increasing BPD severity [10]. Infants with pulmonary interstitial emphysema had also severely abnormal values in the range of 0.15 [11] and sick newborns with congenital diaphragmatic hernia in the early postnatal days had median V_A/Q values of 0.20 with a value less than 0.15 being predictive of subsequent death with good specificity and sensitivity [12].

Interventions to Improve V_A/Q Matching

A small degree of ventilation to perfusion inequality can be attributed to the difference in the gravitational effect on ventilation and perfusion, as blood flow is more gravity dependent and in an upright adult is increased in the basal lung segments compared to the apical ones. Contrarily, ventilation is more uniform across the lung parts and less dependent on gravity. In standing adults, these regional differences give rise to a lower V_A/Q in the basal lung and a higher V_A/Q in the apical lung (lower Q) and a small "physiological" overall mismatch in V_A/Q when both lungs are considered as a single unit [1]. In the neonatal population where upright position is not encountered, the effect of gravity on the different areas of the lung is more prominent when an infant changes position from supine to prone with an overall mild improvement in the prone position compared to the supine [13]. This improvement in V_A/Q might be due to enhanced overall ventilation in the dorsal lung regions compared to the ventral ones in the prone position, by recruitment of the previously atelectatic large dorsal lung areas for ventilation, while the smaller ventral lung become less ventilated. Similarly an incremental stepwise increase in the delivered pressure in preterm infants on continuous positive airway pressure was associated with an improvement in V_A/Q matching as previously atelectactic areas of the lung open and become better ventilated as the external airway pressure is increased [14].

Questions

Question 1: The following are true for ventilation to perfusion relationships:
(a) Abnormalities are only due to ventilation defects
(b) Are abnormal only when the ratio of ventilation to perfusion in low
(c) A value of 0.9 is abnormal
(d) When impaired this can cause significant hypoxaemia

Question 2: A low ventilation to perfusion ratio:
(a) Can occur if perfusion is decreased
(b) Cannot be the result of airway obstruction
(c) Can be the result of decreased alveolar ventilation
(d) Can happen in pure persistent pulmonary hypertension of the newborn without any lung pathology

Question 3: In bronchopulmonary dysplasia:
(a) A low ventilation to perfusion ratio can be overcome by providing a higher fraction of inspired oxygen
(b) The ventilation to perfusion ratio is high
(c) The ventilation to perfusion ratio is always normal
(d) Development of pulmonary interstitial emphysema is associated with an improvement in the ventilation to perfusion ratio

References

1. Wagner PD, Laravuso RB, Uhl RR, West JB. Continuous distributions of ventilation-perfusion ratios in normal subjects breathing air and 100 per cent O2. J Clin Investig. 1974;54(1):54–68. Epub 1974/07/01.10.1172/JCI107750.
2. West JB. Respiratory physiology: the essentials, vol. viii. 9th ed. Philadelphia: Wolters Kluwer Health/Lippincott Williams & Wilkins; 2012. 200 pp.
3. Dao DT, Kamran A, Wilson JM, Sheils CA, Kharasch VS, Mullen MP, et al. Longitudinal analysis of ventilation perfusion mismatch in congenital diaphragmatic hernia survivors. J Pediatr. 2020;219:160-6 e2. Epub 2019/11/11.10.1016/j.jpeds.2019.09.053.

4. Stewart NJ, Smith LJ, Chan HF, Eaden JA, Rajaram S, Swift AJ, et al. Lung MRI with hyperpolarised gases: current & future clinical perspectives. Br J Radiol. 2022;95(1132):20210207. Epub 2021/06/10.10.1259/bjr.20210207.
5. Dyke JP, Voskrebenzev A, Blatt LK, Vogel-Claussen J, Grimm R, Worgall S, et al. Assessment of lung ventilation of premature infants with bronchopulmonary dysplasia at 1.5 Tesla using phase-resolved functional lung magnetic resonance imaging. Pediatr Radiol. 2023;53(6):1076–84. Epub 2023/02/04.10.1007/s00247-023-05598-6.
6. Sapsford DJ, Jones JG. The PIO2 vs. SpO2 diagram: a non-invasive measure of pulmonary oxygen exchange. Eur J Anaesthesiol. 1995;12(4):375–86.
7. Roe PG, Jones JG. Analysis of factors which affect the relationship between inspired oxygen partial pressure and arterial oxygen saturation. Br J Anaesthesia. 1993;71(4):488–94. Epub 1993/10/01.10.1093/bja/71.4.488.
8. Dassios T, Ali K, Rossor T, Greenough A. Using the fetal oxyhaemoglobin dissociation curve to calculate the ventilation/perfusion ratio and right to left shunt in healthy newborn infants. J Clin Monit Comput. 2019;33(3):545–6. https://doi.org/10.1007/s10877-018-0168-6.
9. Dassios T, Curley A, Morley C, Ross-Russell R. Using measurements of shunt and ventilation-to-perfusion ratio to quantify the severity of bronchopulmonary dysplasia. Neonatology. 2015;107(4):283–8. Epub 2015/03/15.10.1159/000376567.
10. Svedenkrans J, Stoecklin B, Jones JG, Doherty DA, Pillow JJ. Physiology and predictors of impaired gas exchange in infants with bronchopulmonary dysplasia. Am J Respir Critic Care Med. 2019; 21.10.1164/rccm.201810-2037OC.
11. Williams E, Dassios T, Clarke P, Chowdhury O, Greenough A. Predictors of outcome of prematurely born infants with pulmonary interstitial emphysema. Acta Paediatr. 2019;108(1):106–11. Epub 2018/05/14.10.1111/apa.14400.
12. Dassios T, Shareef Arattu Thodika FM, Williams E, Davenport M, Nicolaides KH, Greenough A. Ventilation-to-perfusion relationships and right-to-left shunt during neonatal intensive care in infants with congenital diaphragmatic hernia. Pediatr Res. 2022; Epub 2022/03/21.10.1038/s41390-022-02001-2.
13. Barka K, Papachatzi E, Fouzas S, Dimitriou G, Dassios T. Respiratory function in ventilated newborn infants nursed prone and supine. Pediatr Pulmonol. 2025;60(4):e71075. Epub 2025/04/01.10.1002/ppul.71075.
14. Bamat NA, Orians CM, Abbasi S, Morley CJ, Ross Russell R, Panitch HB, et al. Use of ventilation/perfusion mismatch to guide individualised CPAP level selection in preterm infants: a feasibility trial. Arch Dis Child Fetal Neonatal Ed. 2023;108(2):188–93. Epub 2022/09/15.10.1136/archdischild-2022-324474.

Oxygen Transport to the Tissues

7

Once oxygen has been transferred into the pulmonary capillaries, it will then be delivered to the peripheral tissues via the systemic circulation. Oxygen has a relatively low solubility in blood, so most of the oxygen in the blood is transferred bound to haemoglobin and a very small fraction is dissolved unbound in the plasma. The detailed description of the biochemical structure and binding sites of haemoglobin are beyond the scope of this chapter, but the relationship of the saturation of haemoglobin with oxygen with the partial arterial pressure of oxygen is described by the oxygen - haemoglobin or oxyhaemoglobin dissociation curve (Fig. 7.1). This curve is sigmoid with a flattened part at higher values of the arterial oxygen tension. Oxygen at high concentrations is toxic for any individual but particularly so for the premature infants who largely lack efficient antioxidant defences. The flattened shape of the curve can thus explain why we aim to achieve less than 100% transcutaneous saturation (SpO_2) values in clinical practice. By achieving saturations close to 100% we cannot be certain on the exact location of the PaO_2 on the x axis and we are not certain that we are not providing excessively high oxygen concentrations, which are toxic [1].

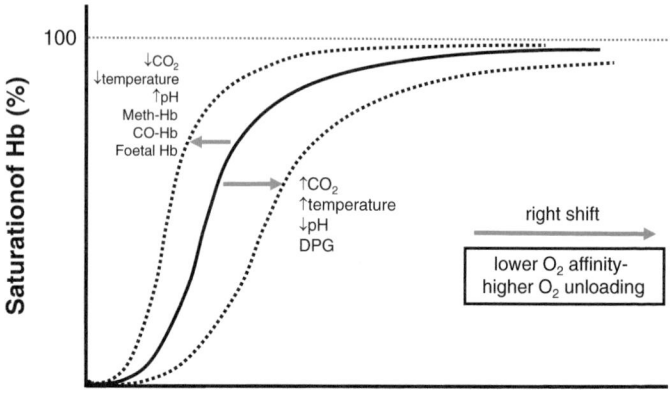

Fig. 7.1 The oxygen haemoglobin dissociation curve. The curve is shifted to the right by higher CO_2, temperature and 2,3-diphosphoglycerate (DPG) and to the left by a higher fraction of meth-haemoglobin (Meth-Hb)), carboxy-haemoglobin (CO-Hb) and foetal haemoglobin. Right shifting of the curve signifies lower affinity of haemoglobin with oxygen and unloading of oxygen from haemoglobin to the peripheral tissues

The Oxyhaemoglobin Dissociation Curve

In clinical practice, diseased lungs will be characterised by a significant amount of intrapulmonary shunting and cannot achieve saturations of 100% or close to 100%. This happens because in the presence of any right-to-left shunting, of either cardiac or pulmonary origin, increasing the provision of the FiO_2 will not improve the SpO_2 as the shunted blood is not exposed at all to ventilation and cannot increase its oxygen content. The degree of the decrease thus of the maximal achievable SpO_2 at high FiO_2 values describes the level of right-to-left shunting which can either be attributed to cardiac or intrapulmonary right-to-left shunting [2, 3].

As described in the previous chapter, the degree of right shift of the ODC relative to an ideal healthy state, is an index of ventilation to perfusion mismatch as supplemental oxygen can remedy the hypoxia which is attributable to impaired ventilation and a low V_A/Q [4, 5]. Displacement of the ODC to the left signifies higher affinity to oxygen and to the right, lower affinity with oxygen. The curve can be shifted to the right by increased temperature, increased concentration of hydrogen ions (acidosis), increased 2,3-diphosphoglycerate and increased values of carbon dioxide (Fig. 7.1). Clinically, some intended deviation from normal body temperature can be seen during whole body hypothermia for hypoxic ischemic encephalopathy. Hypothermia will shift the curve to the left and improve binding of oxygen to haemoglobin and thus improve oxygenation [6].

It is also common clinical practice to tolerate relatively high values of carbon dioxide in ventilated preterm infants in a strategy called *permissive hypercapnia* aiming to minimize lung injury which would occur with excessive tidal volumes and would cause volutrauma [7]. Higher levels of CO_2, however, will also shift the oxyhaemoglobin dissociation curve to the right in a phenomenon first described by the Danish physiologist Christian Bohr in 1904 (the Bohr effect) [8]. In a study of preterm infants with acute or established lung disease, increased levels of CO_2 were indeed associated with a right shift of the ODC and impaired oxygenation and an increase in the right-to-left shunt, possibly as a result of the effect of both hypoxia and hypercapnia in raising pulmonary vascular resistance and thus the pulmonary artery pressure [9]. Both the right shift of the curve and the right-to-left shunt will manifest with a displacement of the ODC to the right due to V_A/Q mismatch and downwards due to the intrapulmonary shunt (Fig. 7.2). Interestingly, despite the fact that permissive hypercapnia is widely adopted as an approach to mitigate harm, at an epidemiological level the clinical benefits of this strategy have not been proven, as permissive hypercapnia has not been associated with a reduction in respiratory or any other morbidity [7, 10].

Another way to measure V_A/Q mismatch and respiratory disease severity with the ODC is by using the anchor point of P_{50}, or the partial arterial pressure of oxygen which corresponds to 50%

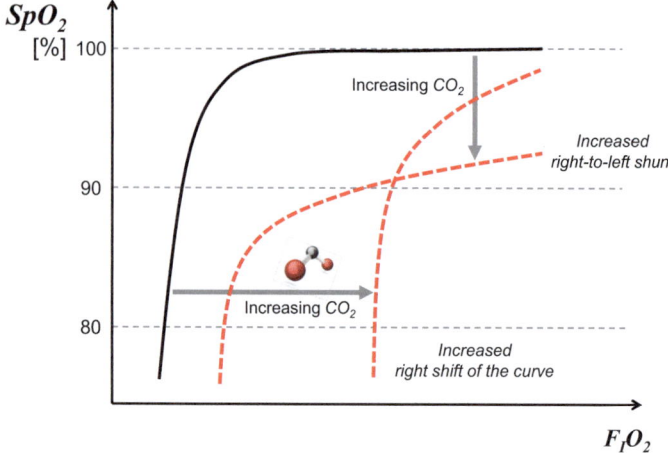

Fig. 7.2 Carbon dioxide and the oxyhaemoglobin dissociation curve. Increased partial pressure of carbon dioxide displaced the curve to the right and caused a downward depression of the curve

saturation of haemoglobin with oxygen. Diseased lungs with V_A/Q mismatch will demonstrate higher values of P_{50} as a higher concentration of inspired oxygen will be required to achieve 50% saturation of oxygen [11]. Higher values of P_{50} describe decreased affinity of hemoglobin for oxygen, and a decreased release of oxygen from the microcirculation to the tissues, and possible tissue hypoxia and subsequent organ dysfunction (Fig. 7.3) [11].

The shape of the ODC can also play a role in an infant's clinical stability as an infant with a very steep curve even with relatively milder lung disease can be stable with minimal oxygen requirements but the same infant might desaturate markedly when oxygen is decreased below a critical level or discontinued. Contrarily, infants with less steep curves which usually correspond to higher SpO_2 values, produce less dramatic changes in SpO_2 in response to changes in the provided oxygen (Fig. 7.4) [12].

The Oxyhaemoglobin Dissociation Curve

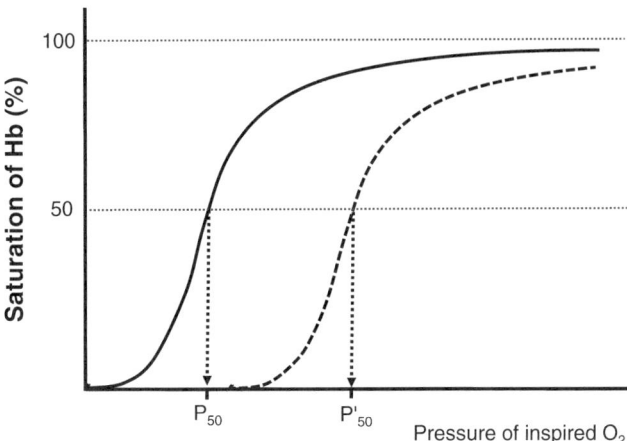

Fig. 7.3 The P_{50} is the oxygen tension at which hemoglobin is 50% saturated with oxygen. When the haemoglobin-oxygen affinity decreases, the oxyhaemoglobin dissociation curve shifts to the right and the P_{50} increases

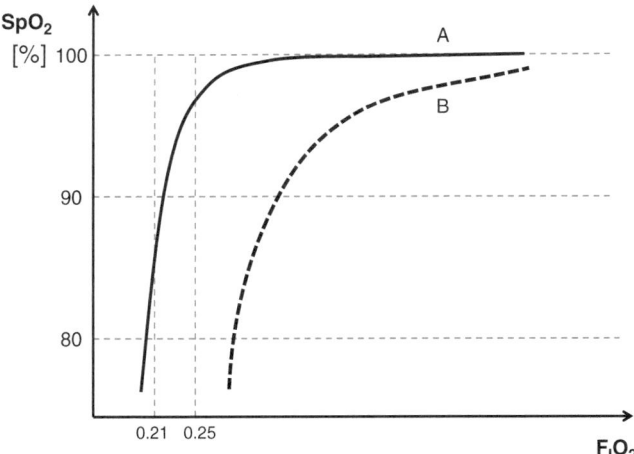

Fig. 7.4 Influence of the slope of the oxyhaemoglobin dissociation curve on clinical stability. Curve A exhibits a steep part where small changes in the provided fraction of inspired oxygen (FiO_2) from 0.25 to 0.21 correspond to a marked decrease in the transcutaneous oxygen saturation (SpO_2) from the high nineties down to the mid eighties. The curve B, although more abnormal and shifted to the right, demonstrates a less steep pattern

Haemoglobin Subtypes

While the hospital laboratory or the gas machine of the Neonatal Unit will provide a single value for the total concentration of haemoglobin in the blood, it is worth remembering that this value is an aggregate of different haemoglobin subtypes. At birth, the haemoglobin is nearly all *foetal* (HbF), with minimal or no adult haemoglobin A (HbA). In utero, HbF is vital for adequate oxygen delivery to the developing foetus as HbF has a higher affinity to oxygen, facilitating thus the transfer of oxygen from the maternal circulation to the fetus. This means that an HbF curve would be positioned to the left compared to an adult or paediatric HbA ODC. Following birth, HbF rapidly declines as the hypoxic stimulus for infants to produce erythropoietin is abolished and in admitted infants repeated blood sampling and blood transfusions replace HbF with HbA. Of note, extremely preterm infants demonstrate a biphasic decline in HbF as the most immature infants demonstrate a second wave of HbF production before HbF levels decline again postnatally, in a phenomenon which is more pronounced in the most immature infants especially the ones born before 24 weeks of gestation (Fig. 7.5) We have previously suggested that it is possible that this phenomenon contributes to the recovery of these infants as the higher affinity of HbF with oxygen might partly explain the decreasing oxygen needs during recovery [13]. The production of HbA is believed to only commence at 30 weeks of gestation, so postnatal HbF production in these infants may persist for longer following birth [14]. In this sense, the observed second peak in the percentage of HbF in infants born before 24 weeks of gestation may be due to a continued postnatal production of foetal haemoglobin.

Another haemoglobin variant of interest is *carboxy-Hb* (COHb) which reflects the endogenous production of carbon monoxide and can be used as a marker of oxidative stress. Although COHb is a very small fraction of the total Hb (usually less than 2%), prematurely born infants who develop disorders of an oxidative background, such as bronchopulmonary dysplasia and intraventricular haemorrhage, have significantly higher COHb levels compared to similarly preterm infants who did not

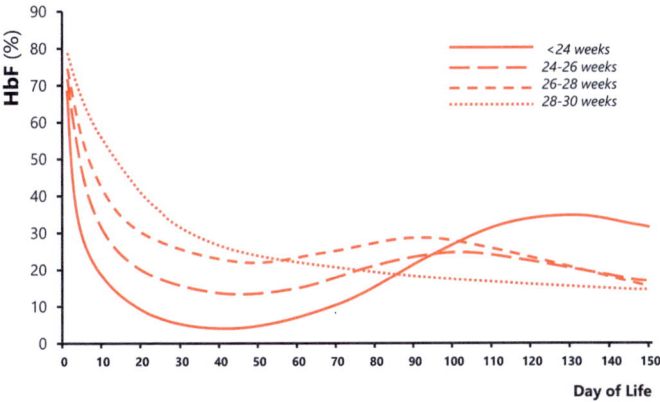

Fig. 7.5 Temporal changes in foetal haemoglobin in extremely preterm infants. Differences in the trends of HbF during the first 150 days of life according to gestational age. The approximate values of HbF in term infants are depicted as solid triangles

develop these complications [15]. Higher levels of COHb have also been reported in hypoxic ischemic encephalopathy [16], signalling a strong pathophysiological interaction between hypoxia and oxidative damage in these infants [17]. Hypoxia and subsequent reoxygenation generate free radicals which could overcome the immature antioxidant defense in the preterm infant leading to oxidative stress and functional and structural tissue damage [18].

The fraction of *methaemoglobin* (MethHb) is also a clinically relevant non-oxygen binding Hb variant. Methaemoglobin is a subtype of haemoglobin in which the ferrous iron has been converted to ferric, which cannot reversibly combine with oxygen, and thus MethHb cannot transport oxygen. In normal subjects MethHb levels are below 2%, but higher levels could be encountered in neonatal clinical care, usually during administration of inhaled nitric oxide [19] and when using glyceryl trinitrate patches for the treatment of digital and limb ischemia [20]. The enzyme erythrocyte b5 reductase is responsible for the reduction of methaemoglobin to haemoglobin and the enzyme activity in the newborn is approximately half that of the adult activity which contributes to the increased susceptibility of infants to methaemoglobinaemia [21].

Carbon Dioxide

Carbon dioxide is more soluble in the blood compared to oxygen so a larger fraction is dissolved in the blood, while the largest part of CO_2 can be found in the form of bicarbonate. When the oxygen saturation decreases at the level of the tissues, deoxygenated haemoglobin has a higher affinity for CO_2, facilitating this way the uptake of CO_2 from the tissues into the peripheral capillaries and the venous circulation. This phenomenon which describes the increased capacity of blood to carry CO_2 under conditions of decreased haemoglobin oxygen saturation is called the Haldane effect.

Questions

Question 1: The oxyhaemoglobin dissociation curve of a specific individual can be shifted to the right:
 (a) When there is more foetal haemoglobin present
 (b) When the partial pressure of carbon dioxide decreases
 (c) During hypothermia
 (d) By acidosis

Question 2: Hypercapnia in the newborn:
 (a) Has been unequivocally proven to reduce respiratory morbidity
 (b) Does not affect oxygenation
 (c) Can depress the oxyhaemoglobin dissociation curve downwards signalling intrapulmonary shunting
 (d) Is commonly used with a view to minimise ventilator-induced lung injury

Question 3: Regarding the haemoglobin subtypes:
 (a) Haemoglobin F has a lower affinity with oxygen
 (b) Haemoglobin F facilitates in utero transfer of oxygen to the foetus from the placenta
 (c) Carboxy-haemoglobin decreases in oxidative disorders
 (d) Meth-haemoglobin must be monitored to be more than 2% during treatment with inhaled nitric oxide

References

1. Tipple TE, Ambalavanan N. Oxygen toxicity in the neonate: thinking beyond the balance. Clin Perinatol. 2019;46(3):435–47. Epub 2019/07/28.10.1016/j.clp.2019.05.001.
2. Roe PG, Jones JG. Analysis of factors which affect the relationship between inspired oxygen partial pressure and arterial oxygen saturation. Br J Anaesth. 1993;71(4):488–94. Epub 1993/10/01.10.1093/bja/71.4.488.
3. Sapsford DJ, Jones JG. The PIO2 vs. SpO2 diagram: a non-invasive measure of pulmonary oxygen exchange. Eur J Anaesthesiol. 1995;12(4):375–86.
4. de Waal K, Kluckow M. Prolonged rupture of membranes and pulmonary hypoplasia in very preterm infants: pathophysiology and guided treatment. J Pediatr. 2015;166(5):1113–20. Epub 2015/02/15.10.1016/j.jpeds.2015.01.015.
5. Wu CS, Chen CM, Chou HC. Pulmonary hypoplasia induced by oligohydramnios: findings from animal models and a population-based study. Pediatr Neonatol. 2017;58(1):3–7. Epub 2016/06/22.10.1016/j.pedneo.2016.04.001.
6. Dassios T, Austin T. Respiratory function parameters in ventilated newborn infants undergoing whole body hypothermia. Acta Paediatr. 2014;103(2):157–61. https://doi.org/10.1111/apa.12476.
7. Wong SK, Chim M, Allen J, Butler A, Tyrrell J, Hurley T, et al. Carbon dioxide levels in neonates: what are safe parameters? Pediatr Res. 2022;91(5):1049–56. Epub 2021/07/08.10.1038/s41390-021-01473-y
8. Bohr C, Hasselbalch K, Krogh A. Über einen in biologischer Beziehung wichtigen Einfluss, den die Kohlensäurespannung des Blutes auf dessen Sauerstoffbindung übt. Skand Arch Physiol. 1904;16:401–12.
9. Dassios T, Williams EE, Kaltsogianni O, Greenough A. Permissive hypercapnia and oxygenation impairment in premature ventilated infants. Respir Physiol Neurobiol. 2023;317:104144. Epub 2023/08/31.10.1016/j.resp.2023.104144.
10. Woodgate PG, Davies MW. Permissive hypercapnia for the prevention of morbidity and mortality in mechanically ventilated newborn infants. Cochrane Datab Syst Rev. 2001;2001(2):CD002061. Epub 2001/06/19.10.1002/14651858.CD002061.
11. Kim Y, Jung JH, Kim GE, Park M, Lee M, Kim SY, et al. P50 implies adverse clinical outcomes in pediatric acute respiratory distress syndrome by reflecting extrapulmonary organ dysfunction. Sci Rep. 2022;12(1):13666. Epub 2022/08/12.10.1038/s41598-022-18038-6.
12. Jones JG, Lockwood GG, Fung N, Lasenby J, Ross-Russell RI, Quine D, et al. Influence of pulmonary factors on pulse oximeter saturation in preterm infants. Arch Dis Child Fetal Neonatal Ed. 2016;101(4):F319–22. https://doi.org/10.1136/archdischild-2015-308675.

13. Bednarczuk N, Williams EE, Kaltsogianni O, Greenough A, Dassios T. Postnatal temporal changes of foetal haemoglobin in prematurely born infants. Acta Paediatr. 2022;111(7):1338–40. Epub 2022/04/17.10.1111/apa.16360.
14. Hellstrom W, Martinsson T, Hellstrom A, Morsing E, Ley D. Fetal haemoglobin and bronchopulmonary dysplasia in neonates: an observational study. Arch Dis Child Fetal Neonatal Ed. 2021;106(1):88–92. Epub 2020/08/28.10.1136/archdischild-2020-319181.
15. Bednarczuk N, Williams EE, Greenough A, Dassios T. Carboxyhaemoglobin levels and free-radical-related diseases in prematurely born infants. Early Human Dev. 2022;164:105523. Epub 2021/12/18.10.1016/j.earlhumdev.2021.105523.
16. Jenkinson A, Zaidi S, Bhat R, Greenough A, Dassios T. Carboxyhaemoglobin levels in infants with hypoxic ischaemic encephalopathy. J Perinatal Med. 2023;51(9):1225–8. Epub 2023/08/28.10.1515/jpm-2023-0174.
17. Kaltsogianni O, Zaidi S, Bhat R, Greenough A, Dassios T. Race, hypoxaemia and oxidative stress in prematurely-born infants. Early Human Dev. 2023;182:105778. Epub 2023/05/02.10.1016/j.earlhumdev.2023.105778.
18. Di Fiore JM, Vento M. Intermittent hypoxemia and oxidative stress in preterm infants. Respir Physiol Neurobiol. 2019;266:121–9. Epub 2019/05/18.10.1016/j.resp.2019.05.006.
19. Hamon I, Gauthier-Moulinier H, Grelet-Dessioux E, Storme L, Fresson J, Hascoet JM. Methaemoglobinaemia risk factors with inhaled nitric oxide therapy in newborn infants. Acta Paediatr. 2010;99(10):1467–73. Epub 2010/05/12.10.1111/j.1651-2227.2010.01854.x.
20. Mintoft A, Williams E, Harris C, Kennea N, Greenough A. Methemoglobinemia during the use of glyceryl trinitrate patches in neonates: two case reports. AJP Rep. 2018;8(4):e227–e9. Epub 2018/10/23.10.1055/s-0038-1669945.
21. Eng LI, Loo M, Fah FK. Diaphroase activity and variants in normal adults and newborns. Br J Haematol. 1972;23(4):419–25. Epub 1972/10/01.10.1111/j.1365-2141.1972.tb07076.x.

Mechanics of Breathing

8

Basic pulmonary mechanics include indices such as the compliance of the respiratory system, the resistance of the respiratory system and the inspiratory time constant. Both the resistance and the compliance of the respiratory system can be measured by inbuilt sensors of the ventilator and can be displayed on the ventilator screen. The flow sensor measures the flow, which can be integrated over time to present the delivered tidal volume, while a proximal pressure sensor can measure the pressure changes at the level of the endotracheal tube. These indices can be calculated and displayed on the ventilator home screen at an inflation-by-inflation basis.

Compliance

Compliance is a measure of how stiff the lungs are, i.e. how much pressure is required to achieve a certain amount of volume change. It is defined as the change in volume that occurs per unit change in the applied pressure. Lung compliance can be calculated by the change in lung volume divided by the change in transpulmonary pressure [1]. Compliance is conceptually a positive/desirable parameter. We usually consider the lungs to be compliant, meaning that they comply well (they expand) with the small pressures that we apply on them. The inverse of compliance (1/C) is called elastance. A high compliance is usually clinically advantageous, while a low compliance is a negative index as non-compliant lungs (or stiff lungs) need a lot of pressure to achieve the desired volume change which is required for gas exchange.

It is interesting to note that in the context of the respiratory distress syndrome (RDS) which is the most common neonatal respiratory disease, lungs with a low compliance were traditionally informally described as "stiff lungs" implying that they are hard to ventilate when using manual techniques such as a bag-valve mask, while lungs with high compliance are referred to as "soft lungs" for the opposite reasons. Radiographically, high-compliance, soft lungs would appear with black lung fields on the chest radiograph, while low-compliance, stiff lungs will appear as opacified, ground-glass lung fields (white or grey on the radiograph).

Typically a low compliance state in neonatology is the initial phase of RDS in preterm infants where one might start with very low compliance values, in the area of 0.2–0.3 ml/cm H_2O which can rapidly increase after administration of surfactant. This rapid change of compliance following administration of surfactant is one of the main reasons that negate the use of volume-targeted ventilation especially in the preterm infants.

If mechanical ventilation in RDS was delivered in a pressure-controlled mode, then the ventilator would deliver steadily a constant distending pressure, for example 20 cm H_2O in a one kg infant, and in the early phase of RDS before the administration of surfactant this would lead to a volume change of 5 ml. Let's suppose that after the administration of surfactant the compliance doubles from 0.3 to 0.6 ml/cm H_2O. This doubling of compliance, which is a realistic scenario in clinical practice, would lead to a doubling of the delivered volume to 10 ml (or 10 ml/kg) in this infant, which would be excessive. Excessive inadvertent delivery of a high tidal volume in a ventilated preterm infant constitutes an injurious process (causing volutrauma via stretching of the epithelial cells) and initiates the cascade of inflammation, which might lead to a chronic inflammatory state and contribute significantly to the development of chronic respiratory morbidity and bronchopulmonary dysplasia (BPD) [2]. If however, the same infant was managed on volume-targeted ventilation, despite the change in compliance following surfactant, the delivered volume would remain constant at 5 ml while the applied pressure would decrease to avoid overstretching of the lung parenchyma and volutrauma would thus had been largely avoided.

This concept has been clearly demonstrated by some animal studies, where preterm rats were ventilated with high pressures, but in a subgroup the volutrauma was mitigated by strapping the chest with a plastic tape to avoid overexpansion, while another group was ventilated with equally high pressures but no chest strapping, so that the lungs could freely over-expand and be exposed to volutrauma. The rats with the strapped chest exhibited minimal lung damage, while the ones without the strap had significant and severe lung injury secondary to these high-pressure inflations. This experiment demonstrates that lung injury is initiated by excessive volume delivery, and this is why we should limit volume extremes and monitor the tidal volumes, rather than the required pressures to achieve these volumes [3, 4]. At an epidemiological level, volume-targeted ventilation has also been shown to decrease chronic respiratory morbidity in the form of BPD, presumably based on the mechanical phenomena highlighted above [5].

Compliance Sigmoid Curve

While compliance of the respiratory system is an index of pulmonary mechanics which can be derived at a breath-by-breath basis, it is worth knowing that from a mechanical perspective, the compliance of the respiratory system as a whole includes values which might not be used in a specific individual at a specific time. If one could plot all the values of pressure and volume from 0 to the maximum values, the derived curve would follow a sigmoid pattern as demonstrated in Fig. 8.1. This sigmoid curve has two flat parts, at the lower and higher range of the administered pressure values. This means that when an infant is ventilated at very low pressures (for example from 0 to 5 cm H_2O), using a set pressure differential (5 cm H_2O) then the achieved volume change would be relatively small. This is ventilation in the *zone of atelectasis*,

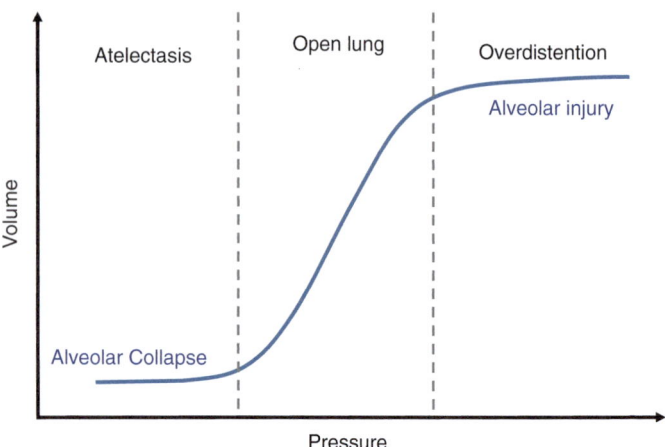

Fig. 8.1 Compliance of the respiratory system. Pressure—Volume curve demonstrating that the volume change is maximised for a given pressure change and ventilation is optimised in the open lung zone which avoids the extremes of atelectasis or overinflation

where high pressures are needed to open the lungs because of repeated alveolar collapse and atelectasis. Similarly, when ventilation is delivered at very high pressures from 15 to 20 cm H_2O and with the same pressure differential (5 cm H_2O), then equally, little volume change can be achieved. This is ventilation in the *zone of overdistention*, and is equally inefficient as a lot of energy is applied in an already over-expanded lung which cannot expand much more and this excessive pressure can cause parenchymal lung injury. In between these two zones, lies the "*open lung zone*", where by applying the same pressure differential of 5 cm H_2O—now from 5 to 10 cmH_2O, considerably higher volume changes can be achieved.

The practical application of this diagram is that we should aim to ventilate in this open lung zone, and this can be achieved by setting our positive end expiratory pressure (PEEP) at the point where the lungs open and remain open, and use the minimum required pressures above this to achieve a certain volume change with each breath (which is usually in the area of 5 ml/kg).

Lung Recruitment Manoeuvres

The compliance curve (or pressure-volume curve) is also useful for clinically selecting the opening pressure that is set at initiation of high frequency oscillation. It is useful to remember that because of the differences in the mechanical properties of the lung at inspiration and expiration the pressure needed for a collapsed lung to open, is not the same as the pressure needed for an open lung to remain open. The first is the *opening pressure* (the pressure that is required to open up the previously atelectactic areas of a collapsed lung) and the second is just above the *closing pressure* (the pressure at which a previously open lung starts to close down again, i.e. the alveoli start to collapse). The opening pressure is higher than the closing pressure and this makes sense at a purely mechanical level, as one specific closed space or cavity requires a higher pressure to open up, compared to the pressure needed merely to keep it open. This is similar to blowing in a balloon, where at the initial phases one needs to blow harder to start inflating the balloon, while if one just needs to keep it open, much less effort would be required.

This has a direct practical application in setting the Mean Airway Pressure (MAP) when switching a ventilated infant with diseased (closed) lungs from conventional to high frequency oscillation. One can start by gradually and carefully increasing the MAP in a step-wise fashion until the point of maximum and rapid recruitment is reached, which will manifest clinically with a sudden increase of the saturation of oxygen as previously atelectatic lung segments start to open up (opening pressure). Then, when recruitment has been achieved, and to avoid the provision of excessive and injurious pressures, the MAP is gradually titrated down to the point when saturation starts to decrease as the lungs start to close again (closing pressure). At this point, the operator can go up by one or two cm H_2O to ensure the lung stays open and the lungs can then be ventilated using this "optimal" pressure (Fig. 8.2) [6].

This manoeuvre is demonstrated in ventilation workshops by using an animal lung preparation, where one can clearly see mac-

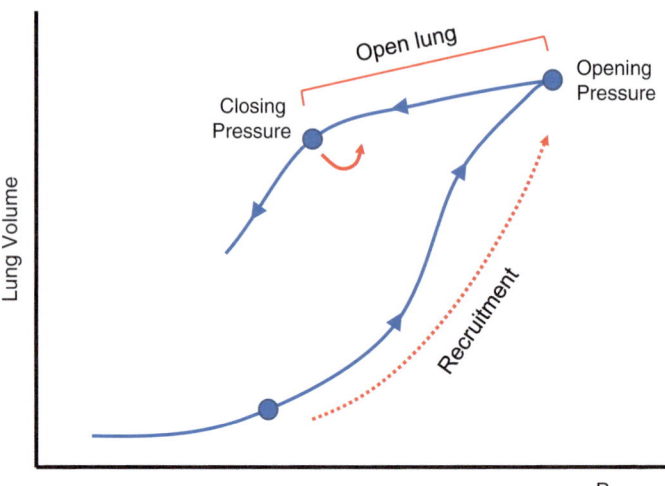

Fig. 8.2 Lung recruitment manoeuvre. Diagrammatic representation of the airway pressure (P_{aw})—volume relationship in an atelectatic diseased lung. Active lung recruitment occurs starting from the starting point along the inflation limb of the PV curve as P_{aw} is step-wise increased to the opening pressure where the lung volume approaches total lung capacity or overdistension. Reductions in P_{aw} from the opening pressure maintain lung recruitment along the deflation limb of the PV curve until the closing pressure where derecruitment starts to occur. The optimal P_{aw} for lung recruitment and oxygenation occurs 1–2 cm H_2O above the closing pressure

roscopically by visually assessing the lung volumes, that the pressure required to open up a previously closed lung is always higher than the minimum pressure needed to keep it open. As the pressure volume curve is usually displayed on the ventilator screen, some further applications of respiratory monitoring via this curve will be discussed in the chapter on "Ventilator waveforms".

Resistance

Resistance is the other major index of respiratory mechanics which is measured by the ventilator and can be displayed on real time on the ventilator screen.

Resistance

While compliance is an index that primarily describes the lung parenchyma, resistance primarily refers to airway function. Airway resistance is the resistance that occurs between moving molecules of gas and between these molecules and the wall of the respiratory system, such as the trachea, the bronchi and the bronchioles. A high resistance in a ventilated subject can either be due to purely iatrogenic mechanical parameters related to the apparatus used such as the ventilator circuit, endotracheal tube and measuring sensors or in can be due to real clinical pathology affecting the neonatal airways.

The resistance of flow through a tube can be calculated by the law of Poiseuille according to the following formula:

$$R = 8vL / \pi r^4$$

where R is resistance, v is viscosity, L is length of the tube (airway), and r is the radius of the tube.

It is evident that since the radius is on the fourth power, the diameter of the endotracheal tube will have a considerably larger impact on the resistance than the length of the tube, and this is why in clinical practice when aiming to decrease the resistance in a ventilated subject, it is much more efficient to use a wider endotracheal tube if possible, than cutting the endotracheal tube short to decrease the length. It is important to remember that the resistance of flow across a system is determined by the narrowest part of that system, which in our case is the endotracheal tube. One might think that subdivisions on the bronchial tree below the trachea would give rise to narrower airways than the endotracheal tube, so the resistance there would be higher. This is true at an individual tube/airway level, but the multiple divisions of the airways going down to the level of the alveoli, produce a progressively increasing *total* number of airways and a larger total cross sectional area which resembles the shape of a trumpet (Fig. 1.4). It is this *total* cross sectional area which will determine the resistance, because the aggregate of the smaller airways is larger than the single wider trachea or endotracheal tube. The most narrow point of the respiratory system, corresponding to the higher airway resistance, thus in a healthy ventilated subject is the endotracheal tube, and if that infant is not ventilated it is the larynx and the trachea.

Increased Airway Resistance

The clinical significance of increased airway resistance in an individual with respiratory disease is that very high resistance might impede the flow of gases to the gas-exchanging parts of the lung especially when higher flows are used. When a gas is delivered via a long and narrow airway, it takes longer for it to reach the end of the airway and the flow becomes turbulent rather than laminar, which further impacts the delivery of gas (Fig. 8.3). As with most airway-related pathologies, this phenomenon is more pronounced during expiration, where increased resistance becomes more prominent and more problematic as the airways narrow down during expiration due to airway dynamic compression. This is why all high-airway resistance disorders first become clinically apparent at expiration, and when they progress to become more severe, they go on to affect inspiration as well. For example, mild asthma will clinically start with a prolonged *expiratory* phase, it will then manifest with *expiratory* wheezing, and if the stenosis becomes more severe, *inspiratory* wheezing will also appear.

In neonatal critical care, increased airway resistance very commonly occurs in ventilated infants secondary to impacted material in the endotracheal tube, or the trachea and the large airways such as mucous plugs, secretions or blood clots. Active suctioning and removal of these materials could acutely decrease resistance and facilitate the delivery of the ventilator gases. Increased resistance secondary to evolving pulmonary disease typically occurs in established BPD and is fixed in nature because of the structural stenosis of the airways which is characteristic of BPD. This prac-

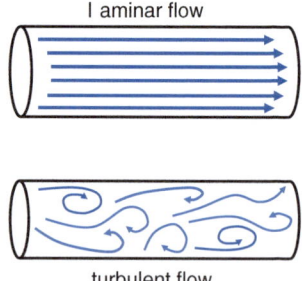

Fig. 8.3 Laminar and turbulent flow. The flow of gas through a stenotic airway becomes turbulent impacting the delivery of the gas to the alveoli

tically means that when such an infant is invasively ventilated a longer inspiratory and expiratory time will need to be provided to achieve adequate inflation and full deflation of the lung. While newborns in the first days of life are typically ventilated with a short inspiratory time in the area of 0.3–0.4 s and a total rate of 40 or 50 per minute, a ventilated infant with severe BPD might require an inflation time in excess of 0.8 s which will negate a slower rate of approximately 20 per minute to allow for full deflation of the lung, as it is the expiratory phase which is mostly affected by increased airway resistance.

Another condition in neonatal medicine where high airway resistance might be encountered is *meconium aspiration syndrome* where thick articulate meconium lines and obstructs the airways and increases the resistance to the movement of gas through these airways. Similarly, longer inflation and deflation times might be required. It is worth highlighting the presence of the *ball-valve phenomenon* which occurs during mechanical ventilation of infants with meconium aspiration syndrome, where the meconium obstruction can be overcome by higher pressures during inflation but the same airways become occluded during deflation as they narrow down and this can lead to gas trapping peripherally to the meconium and an increased occurrence of air leaks such as pneumothorax. For these reasons, high frequency oscillation is also a sensible ventilatory approach in meconium aspiration syndrome as the lungs are kept open with a constant pressure and the ball-valve effect is avoided [7].

Flow-Volume Curves

Airway resistance in ventilated subjects can be monitored and visualised with the flow versus time and the flow versus volume curves. Increased resistance in the flow versus time curve will manifest with a delay or a failure of the expiratory flow to return back to zero during expiration (Fig. 8.4) and with a steep expiratory part of the flow volume curve demonstrating increased expiratory resistance. These curves can be utilised to fine tune the ventilator parameters so that these problems can be mitigated. For example, in a ventilated infant with severely increased expiratory

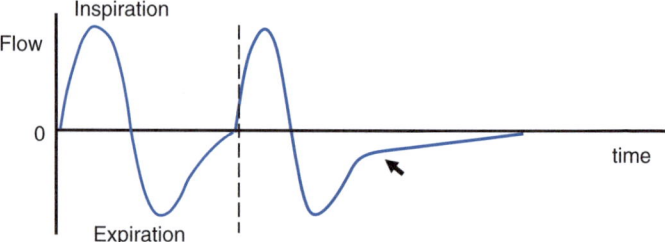

Fig. 8.4 Flow versus time curve. The first part shows a normal inspiratory and expiratory flow signal and the second part a curve of a subject with increased expiratory resistance where the flow signal takes longer to return to zero

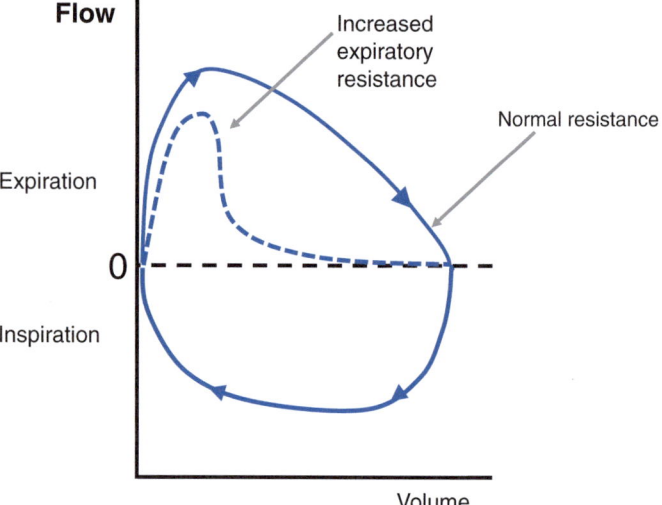

Fig. 8.5 Flow volume curve. Increased expiratory resistance compared to a normal subject

resistance secondary to tracheomalacia and collapse of the tracheal walls during expiration, titration of the positive end expiratory pressure by providing higher pressures, might reach the point where the trachea remains open at end expiration, which visually will be the point at which the expiratory part of the flow volume curve remains round and is no longer steep (Fig. 8.5). Some more examples of the clinical utility of the flow volume loops will also be presented in the chapter on the Ventilator waveforms.

Inspiratory Time constant

The product of the compliance times the resistance equals with the inspiratory time constant.

Conceptually, the time constant in a ventilated subject describes how long it takes for the pressure provided at the level of the endotracheal tube to be delivered to the lungs and this time is influenced by the mechanical properties of the lung.

As a mechanical analogue, it might be useful to think of blowing into a stiff, wide and short straw. Because the straw is stiff (low compliance) wide and short (low resistance) whatever pressure we blow into it, that pressure will very quickly transfer to the other end of the straw. This is the equivalent of a small time constant. If, however, we blew with the same effort in a very soft and long balloon, it will take a longer time for the pressure that we apply at the opening to reach the other end of the balloon (long time constant) (Fig. 8.6). Practically, this means that if we are ventilating lungs with a low compliance and normal resistance

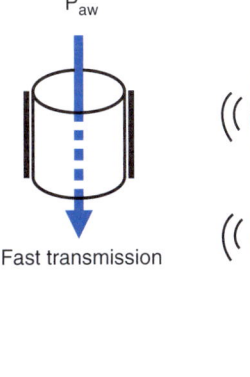

Fig. 8.6 Inspiratory time constant. The flow of gas through a short airway surrounded by non-compliant parenchyma is fast (low time constant), while the same gas going through a longer, narrow airway surrounded by compliant parenchyma is slower (increased time constant)

(such as for example in RDS) we do not need a very long inspiratory time to deliver this pressure down to the lungs. If, though, we are ventilating lungs with a very high resistance and maybe high compliance (for example in BPD) then we will require a longer inspiratory time for the pressure to reach the alveoli and deliver the gas at the gas-exchanging membrane.

At the bedside, lungs with a small time constant are sometimes informally referred to as "fast lungs", while lungs with a long time constant are called "slow lungs". Since the time constant can be calculated as the product of compliance times resistance, it is common practice to allow 3-5 time constants for inspiratory time and at least an equal or longer time for expiration [8]. For example in a preterm infant with an inspiratory time constant of 0.05–0.1 s, an inspiratory time of 0.3–0.4 s will most likely be sufficient to adequately inflate the lungs.

In some cases of lung disease which involve increased airway resistance with relatively preserved lung compliance, ventilated infants might be at risk of incomplete emptying of the lungs especially when the applied assisted ventilation does not allow for sufficient time for exhalation, as the lungs have an abnormally long time constant. These situations might result in gas trapping, an increase in the lung volume and a build-up of pressure in the alveoli and distal airways referred to as *inadvertent positive end expiratory pressure* (PEEP) or *auto PEEP* [9].

Finally, although commonly it is useful to think of the respiratory system as a single compartment with a single value for compliance and resistance, lung disease in clinical practice is very often inhomogeneous in distribution and different areas of the respiratory system have differing and often opposing mechanical properties. Such an example of ventilation inhomogeneity is typically seen in BPD where separate units have radiographically inhomogeneous appearance and mechanical properties, and BPD lungs are often characterised by areas of alternating collapse and consolidation or overexpansion. The resistance and compliance values we obtain from pulmonary function measurements and which are presented on the ventilator screen, are essentially weighted means for the respiratory system taken as a theoretical single unit.

Questions

Question 1: The compliance of the respiratory system:
  (a) Is calculated as the ratio of the change in volume divided by the change in pressure
  (b) When it is very low this means the lungs are soft and easy to ventilate
  (c) Is an index that primarily describes airway function
  (d) Can be visually depicted by the pressure versus volume diagram

Question 2: In the respiratory distress syndrome of the newborn:
  (a) Compliance of the respiratory system starts by being high and then decreases with the administration of surfactant
  (b) During pressure-controlled ventilation the ventilator will automatically adapt to the change in compliance by giving lower pressures
  (c) During volume-targeted ventilation the delivered pressure will remain the same so the compliance will not change
  (d) Excessively high tidal volumes can damage the premature lungs

Question 3: The following are true regarding the compliance of the respiratory system during invasive ventilation:
  (a) Ventilating with unnecessary high pressures increases the compliance
  (b) Ventilating at the lowest pressure we can set, always keeps the lungs open
  (c) The positive end expiratory pressure corresponds to the maximum value of the "open lung zone"
  (d) The peak inflating pressure should not exceed the higher value of pressure in the "open lung zone"

Question 4: The following are true regarding the lung recruitment manoeuvre:
  (a) The opening pressure is always lower than the closing pressure
  (b) We should ventilate by using pressures below the opening pressure
  (c) The opening pressure clinically corresponds to the point when optimal oxygen saturation is achieved
  (d) The mean airway pressure in high frequency oscillation should be 5 cm H_2O below the closing pressure

Question 5: The resistance of the respiratory system:
  (a) Is primarily an index of parenchymal lung properties
  (b) Is independent of the radius of the airways
  (c) Is more sensitive to changes in the length of the endotracheal tube rather than the diameter
  (d) Is at its highest at the level of the small peripheral airways
  (e) Is significantly influenced by the size of the endotracheal tube

Question 6: In diseases with increased airway resistance:
  (a) The inspiratory phase will be more severely affected than the expiratory phase
  (b) A longer expiratory phase (or deflation time) might be required
  (c) There can be prolonged expiration
  (d) BPD is not characterised by increased resistance
  (e) Removal of meconium by suction will increase the resistance

Question 7: The following are true for the inspiratory time constant:
  (a) Is the ratio of the compliance of the respiratory system divided by the resistance
  (b) Is useful during mechanical ventilation in setting the positive end expiratory pressure
  (c) Is useful during mechanical ventilation for setting the inspiratory time
  (d) Is increased in respiratory distress syndrome
  (e) Is decreased in bronchopulmonary dysplasia

References

1. Desai JP, Moustarah F. Pulmonary compliance. StatPearls. Treasure Island (FL) ineligible companies. Disclosure: Fady Moustarah declares no relevant financial relationships with ineligible companies. 2025.
2. Jobe AH, Ikegami M. Mechanisms initiating lung injury in the preterm. Early Human Dev. 1998;53(1):81–94. Epub 1999/04/08.10.1016/s0378-3782(98)00045-0.
3. Dreyfuss D, Soler P, Basset G, Saumon G. High inflation pressure pulmonary edema. Respective effects of high airway pressure, high tidal volume, and positive end-expiratory pressure. Am Rev Respir Dis. 1988;137(5):1159–64. Epub 1988/05/01.10.1164/ajrccm/137.5.1159.
4. Dreyfuss D, Saumon G. Ventilator-induced lung injury: lessons from experimental studies. Am J Respir Critic Care Med. 1998;157(1):294–323. Epub 1998/01/28.10.1164/ajrccm.157.1.9604014.
5. Wheeler KI, Klingenberg C, Morley CJ, Davis PG. Volume-targeted versus pressure-limited ventilation for preterm infants: a systematic review and meta-analysis. Neonatology. 2011;100(3):219–27. Epub 2011/06/28.10.1159/000326080.
6. Solis-Garcia G, Gonzalez-Pacheco N, Ramos-Navarro C, Vigil-Vazquez S, Gutierrez-Velez A, Merino-Hernandez A, et al. Lung recruitment in neonatal high-frequency oscillatory ventilation with volume-guarantee. Pediatr Pulmonol. 2022;57(12):3000–8. Epub 2022/08/24.10.1002/ppul.26124.
7. Dargaville PA. Respiratory support in meconium aspiration syndrome: a practical guide. Int J Pediatr. 2012;2012:965159. Epub 2012/04/21.10.1155/2012/965159.
8. Kamlin C, Davis PG. Long versus short inspiratory times in neonates receiving mechanical ventilation. Cochrane Datab Syst Rev. 2004;2003(4):CD004503. Epub 2004/10/21.10.1002/14651858. CD004503.pub2.
9. Simbruner G. Inadvertent positive end-expiratory pressure in mechanically ventilated newborn infants: detection and effect on lung mechanics and gas exchange. J Pediatr. 1986;108(4):589–95. Epub 1986/04/01.10.1016/s0022-3476(86)80845-9.

Work of Breathing 9

For gas exchange to occur, air needs to be transported in the lungs by tidal breathing. The energy required to perform this action over a predefined amount of time is described by the term *work of breathing*. This work is undertaken by the respiratory muscles. The main muscle of respiration is the diaphragm, which morphologically appears as a dome-shaped thin muscular layer which separates the abdominal from the thoracic cavity. The diaphragm contracts during inspiration and generates the negative intrathoracic pressure which is required for air to be drawn into the chest. Other respiratory muscles that contribute to the work of breathing to a smaller extent are the intercostal, the sternocleidomastoids and the scalenes, while in laboured expiration the abdominal muscles may also be recruited [1].

Practically, a progressive failure of the diaphragm to adequately undertake the work of breathing will initially manifest with signs of respiratory distress like tachypnoea and intercostal recessions and will eventually lead to respiratory failure which will require the initiation of assisted ventilation.

The Neonatal Diaphragm

The diaphragm of the newborn, and especially of the preterm infant, is mechanically disadvantaged to undertake the work of breathing. In older children and adults, elevation of the ribs during inspiration increases the intra-thoracic volume, but the newborn ribs are already horizontal, and their elevation has limited effect on the intra-thoracic volume [2]. The neonatal chest is also non-ossified and prone to distortion [3]. While in older children and adults the diaphragm attains the shape of a higher dome, the neonatal diaphragm is morphologically more flattened and connects to the chest wall more perpendicularly (with a larger angle), resulting in a decreased range of displacement when contracting (Fig. 9.1). In the newborn, the posterior limb of the diaphragm can be distorted when in respiratory distress, a phenomenon which is clinically described as "subcostal recession". At the histological level, the newborn diaphragm has fewer fatigue-resistant slow twitch (type-I) fibres and a decreased oxidative capacity [4]. The infantile thorax is also overall more cylindrical at a transverse plane, compared to the adult thorax which is more ellipsoidal in shape. These anatomical and functional characteristics render the neonatal diaphragm prone to dysfunction and muscle fatigue.

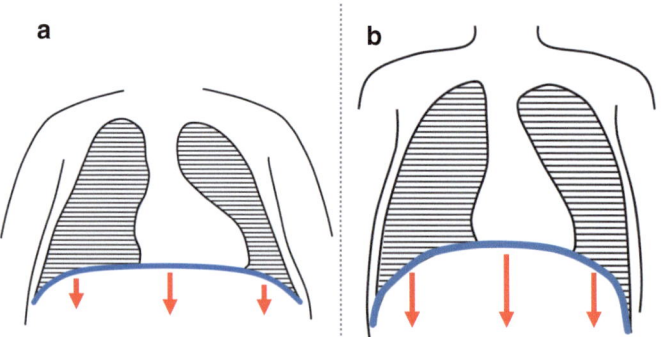

Fig. 9.1 The neonatal (a) and adult (b) diaphragm. Contraction from a more flattened resting state in the neonatal diaphragm diminishes the mechanical efficiency of the diaphragm

Functional Properties of the Respiratory Muscles

As with any skeletal muscle group, the respiratory muscles can be assessed on the basis of their ability to generate force (muscle strength) and their ability to sustain their action over time (muscle endurance) and resist fatigue. For example, the lower limb muscles of a sprint runner can generate a high force in a short period of time without having enhanced endurance properties but in a long-distance runner they are trained to sustain their action over longer periods of time usually at a lower force-generating level. Similarly, for the respiratory muscles both these properties are required for sustained efficient gas exchange and can be separately assessed [5].

Methods to Measure Neonatal Respiratory Muscle Function

A wide variety of methods have been used to assess respiratory muscle function in the newborn and these methods can be used in a targeted way to investigate properties of strength or endurance or a combination of the two. Electromyography can be applied either with surface or orogastric/oesophageal electrodes and detects the electromyographic signal which is produced during the contraction of the diaphragm. Electromyography is a technique that quantifies the electrical activity of the diaphragm and can be used to assess respiratory muscle control, the estimation of the intensity of the respiratory motor output and the efficacy of muscle contraction [5].

Maximal Respiratory Pressures

The maximal inspiratory (P_{Imax}) and maximal expiratory (P_{Emax}) pressures can be recorded as the most negative and most positive pressures respectively, generated during crying against an occluded airway and can be measured at the level of the mouth [6]. Crying is assumed to represent a maximal effort [7]. The mean P_{Imax} during crying in healthy term infants at a mean age of

1.4 years was 118 cm H_2O, while the mean P_{Emax} was 125 cm H_2O and correlated with body weight [8]. In premature infants, maximal airway pressures during crying have been reported to be significantly lower [9]. One relative disadvantage of the assessment of respiratory muscle function by maximal pressures is that it is an indirect and effort-dependent method, and although higher values are usually sufficient to infer adequate muscle strength, low values are not always indicative of respiratory muscle weakness [10].

Thoraco-Abdominal Asynchrony (TAA)

While in healthy individuals there is synchronous inward and outward motion of the chest and abdomen, non-synchronous movement of the two compartments can occur in TAA. This asynchrony can be measured by respiratory inductive plethysmography and quantified by the phase angle between the chest and the abdominal movement, when the relevant displacements are synchronously plotted against time (Fig. 9.2) [11]. An increased phase angle of TAA in indicative of an increased work of breathing [12].

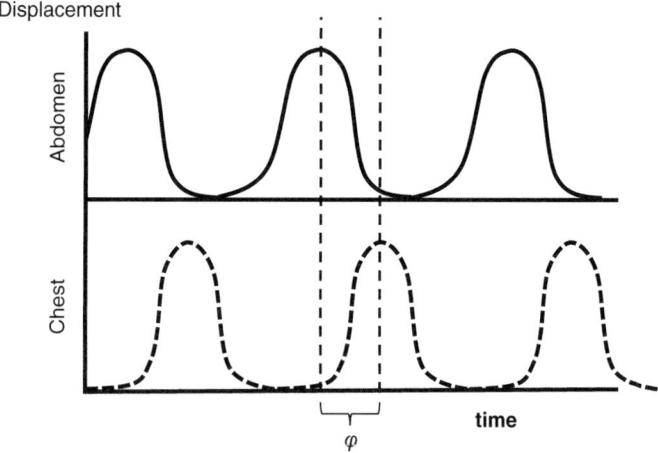

Fig. 9.2 Thoraco-abdominal asynchrony. The chest and abdomen displacements are depicted over time. When the two compartments move in asynchrony, there can be a phase angle (φ) between the two compartments

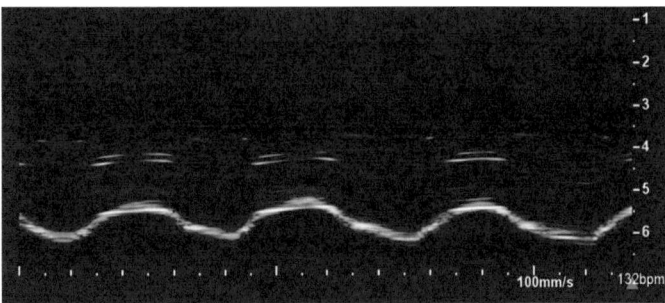

Fig. 9.3 **Diaphragmatic ultrasound**. Sonographic image showing M-mode measurement of the right hemidiaphragm in a healthy term infant

Diaphragmatic Ultrasound

Ultrasonographic assessment of the diaphragm can be undertaken from the right subcostal area, using the liver acoustic window to measure diaphragmatic thickness in the zone of apposition as well as diaphragmatic kinetics by M-mode ultrasonography and recording of the displacement of the diaphragmatic segments and the corresponding velocity (Fig. 9.3) [13].

Prediction of Extubation

Since the work of breathing is undertaken by the respiratory muscles and predominantly by the diaphragm, it follows that indirect confirmation of adequate diaphragmatic muscle function while an infant is still invasively ventilated might have the capacity to predict the ability of that infant to undertake the work of breathing independently when breathing unassisted [14]. Some *composite* indices of respiratory muscle function have been used in this prediction. These are indices that incorporate information both on muscle strength as well as on the ability of the respiratory muscles to undertake the work of breathing over time. Such indices are the *tension time index of the diaphragm*, the *relaxation rate of the inspiratory muscles* and the *electromyographic* signal measured either by indwelling oesophageal and gastric catheters or via adhesive surface electrodes [1]. In premature ventilated infants the predictive ability of these indices is variable and frequently not superior to basic demographic or epidemiologic parameters such as the gestational age and the oxygen requirement at assessment.

Testing of the respiratory muscles prior to extubation is particularly relevant in infants with impaired diaphragmatic function such as infants with congenital diaphragmatic hernia [15] where large diaphragmatic defects would translate to a decreased total diaphragmatic muscle mass and thus a decreased ability to independently undertake the work of breathing [14]. One simplified approach to predict the ability to extubate successfully in premature infants is the spontaneous breathing trial where a ventilated infant is switched to endotracheal continuous positive airway pressure for a brief period of 5–10 min and is observed in their capacity to undertake the work of breathing while breathing unassisted. The test by convention is considered failed if there is bradycardia (heart rate <100 bpm) for over 15 s or a decrease in the transcutaneous oxygen saturation to <85% despite a 15% increase in percentage of the provided oxygen [16, 17].

Factors That Affect Respiratory Muscle Function in the Newborn

At an epidemiological level, low gestation and birth weight are universally and strongly associated with impaired respiratory muscle function, possibly due to decreased muscle mass and functional impairment at lower gestations [18]. The rapid eye movement (REM) phase of sleep is also associated with a loss of intercostal muscle tone and a reduction in post inspiratory muscle activation [19]. Hyperinflation of the chest can place the diaphragm at a mechanical disadvantage as marked hyperinflation would force the diaphragm to flatten and displace caudally and in this position would have a reduced ability to contract any further [20]. This condition is characterised by a decreased force-generating capacity, based on the force–length relationship of the diaphragm, where the optimal length at which the greatest force

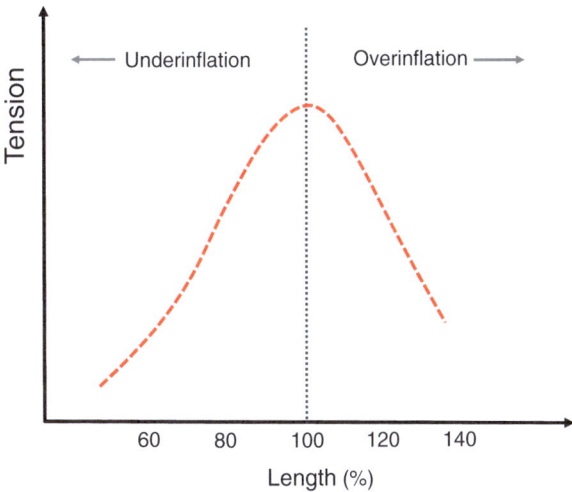

Fig. 9.4 The length-tension diagram of the diaphragm. The maximum tension generated by the diaphragm corresponds to an intermediate resting length which is below the overexpansion zone

output of the diaphragm can be generated, corresponds to a lung volume close to the functional residual capacity (Fig. 9.4) [21].

Diaphragmatic dysfunction can coexist with a haemodynamically significant patent ductus arteriosus. Based on animal experimental and adult studies, the postulated mechanism is that left ventricular heart failure could negatively affect diaphragmatic function via hypoperfusion of the diaphragm and activation of a series of pro-inflammatory cytokines [22]. In ventilated preterm infants this could lead to impaired kinetics of the muscle in affected infants with a lower inspiratory velocity (used here as an index of impeding muscle fatigue) compared to equally premature infants without a significant patent ductus arteriosus [23]. Posture can also affect diaphragmatic function. The prone position can decrease the work of breathing in non-ventilated premature infants compared to the supine position as prone is thought to stabilise the over-compliant neonatal thorax [24]. Studies examining the effect of posture on the work of breathing, assessed by the measurement of the diaphragmatic pressure–time product, in convalescent preterm infants have reported that the work of breathing was lower in the prone compared to the supine position [25].

Some pharmacological agents such as caffeine have been reported to increase diaphragmatic strength [26] and transiently improve the contractility of the diaphragm in the newborn [27], while other agents such as systemic corticosteroids, are associated with universal skeletal muscle atrophy and can negatively affect diaphragm force, weight and endurance [28]. Finally, systemic or respiratory infection can negatively impact respiratory muscle function possibly via generalised spill-over of inflammatory cytokines which could directly affect the diaphragm [29]. In utero inflammation might also impair diaphragmatic function as demonstrated by animal studies where intra-amniotic injection of endotoxin before delivery was associated with impaired diaphragmatic contractility, diaphragmatic proteolysis and muscular atrophy [30].

Neurally Adjusted Ventilatory Assist

Neurally adjusted ventilatory assist (NAVA) is a distinct example of personalised medicine in neonatal ventilation and a model technology highlighting the clinical applications of respiratory muscle physiology. NAVA uses a nasogastric tube with a distal electrode array, which measures the electromyogram of the diaphragm (Edi) measured in microVolts, which is the signal used to trigger the ventilator and determines the level of support in each inflation [31]. During NAVA the level of the delivered respiratory support is proportional to the measured electrical activity of the diaphragm, which is reflective of the neural respiratory drive (Fig. 9.5).

One distinct advantage of NAVA is that the electromyographic signal of the diaphragm during contraction is naturally a superior synchronisation signal compared to flow sensors or other methods, as the contraction of the diaphragm constitutes precisely the early phase of an actual incoming respiratory effort. Furthermore, the provided respiratory support uses the intensity of the electro-

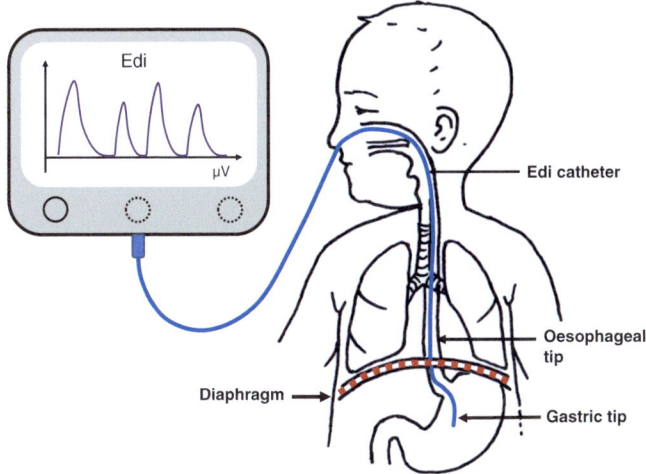

Fig. 9.5 Neurally Adjusted Ventilatory Assist. A nasogastric tube with a distal electrode array (Edi catheter), measures the electromyogram of the diaphragm (Edi) which is the signal used to trigger the ventilator

myographic signal to deliver proportional support at an inflation-to-inflation frequency, eliminating thus any unnecessary excesses of pressure or volume delivery while ensuring adequate support in a process which is regulated in real time. This degree of sophistication seems to be particularly useful in the frailest and most reserve-depleted infants such as extremely preterm infants with chronic respiratory disease which are unable to otherwise wean off invasive ventilation to more conventional modes of non-invasive support [32].

Another potential advantage of NAVA is that it can also be used as a non-invasive modality where it can reduce the reintubation rate after extubation in pre-term neonates [33]. It is also worth remembering that synchronisation via a flow sensor is difficult during non-invasive support, making thus NAVA an advantageous technology for non-invasive ventilation which can detect and augment spontaneous breathing activity.

Questions

Question 1: The neonatal respiratory muscles:
  (a) Are resistant to muscle fatigue compared to older children and adults
  (b) Can generate higher maximum pressures in preterm infants compared to term ones
  (c) Function generally better in preterm infants compared to older children
  (d) Are mechanically disadvantaged to undertake the work of breathing compared to older children

Question 2: In assessing respiratory muscle function in the newborn, the following methods can be used:
  (a) Spirometry
  (b) Maximum respiratory pressures
  (c) Electromyography of the oesophagus
  (d) Lung ultrasound

Question 3: The following can have a **positive** impact on the neonatal respiratory muscles:
  (a) Systemic corticosteroids
  (b) Caffeine citrate
  (c) Supine position
  (d) Extreme prematurity
  (e) Infection

Question 4: During Neurally adjusted ventilatory assist (NAVA):
  (a) The electrical activation of the diaphragm is detected by a flow sensor
  (b) The pressure delivered by the ventilator is always the same
  (c) Is particularly useful in successfully extubating infants with severe BPD
  (d) Can only be used during invasive mechanical ventilation

References

1. Dassios T, Vervenioti A, Dimitriou G. Respiratory muscle function in the newborn: a narrative review. Pediatr Res. 2021; Epub 2021/04/21.10.1038/s41390-021-01529-z.

References

2. Hershenson MB, Colin AA, Wohl ME, Stark AR. Changes in the contribution of the rib cage to tidal breathing during infancy. Am Rev Respir Dis. 1990;141(4 Pt 1):922–5. https://doi.org/10.1164/ajrccm/141.4_Pt_1.922.
3. Papastamelos C, Panitch HB, England SE, Allen JL. Developmental changes in chest wall compliance in infancy and early childhood. J Appl Physiol. 1995;78(1):179–84.
4. Sieck GC, Fournier M, Blanco CE. Diaphragm muscle fatigue resistance during postnatal development. J Appl Physiol. 1991;71(2):458–64.
5. Laveneziana P, Albuquerque A, Aliverti A, Babb T, Barreiro E, Dres M, et al. ERS statement on respiratory muscle testing at rest and during exercise. Eur Respir J. 2019;53(6) Epub 2019/04/09.10.1183/13993003.01214-2018.
6. Gaultier C. Respiratory muscle function in infants. Eur Respir J. 1995;8(1):150–3. https://doi.org/10.1183/09031936.95.08010150.
7. Kosch PC, Stark AR. Dynamic maintenance of end-expiratory lung volume in full-term infants. J Appl Physiol Respir Environ Exerc Physiol. 1984;57(4):1126–33.
8. Shardonofsky FR, Perez-Chada D, Carmuega E, Milic-Emili J. Airway pressures during crying in healthy infants. Pediatr Pulmonol. 1989;6(1):14–8.
9. Dimitriou G, Greenoug A, Dyke H, Rafferty GF. Maximal airway pressures during crying in healthy preterm and term neonates. Early Human Dev. 2000;57(2):149–56.
10. Decramer M, Scano G. Assessment of respiratory muscle function. Eur Respir J. 1994;7(10):1744–5.
11. Hammer J, Newth CJ. Assessment of thoraco-abdominal asynchrony. Paediatr Respir Rev. 2009;10(2):75–80. https://doi.org/10.1016/j.prrv.2009.02.004.
12. ATS/ERS Statement on respiratory muscle testing. Am J Respir Critic Care Med. 2002;166(4):518–624. Epub 2002/08/21.eng.
13. Laing IA, Teele RL, Stark AR. Diaphragmatic movement in newborn infants. J Pediatr. 1988;112(4):638–43.
14. Shalish W, Latremouille S, Papenburg J, Sant'Anna GM. Predictors of extubation readiness in preterm infants: a systematic review and meta-analysis. Arch Dis Child Fetal Neonatal Ed. 2019;104(1):F89–97. https://doi.org/10.1136/archdischild-2017-313878.
15. Dimitriou G, Greenough A, Kavvadia V, Davenport M, Nicolaides KH, Moxham J, et al. Diaphragmatic function in infants with surgically corrected anomalies. Pediatric Res. 2003;54(4):502–8. https://doi.org/10.1203/01.PDR.0000081299.22005.F0.
16. Williams EE, Arattu Thodika FMS, Chappelow I, Chapman-Hatchett N, Dassios T, Greenough A. Diaphragmatic electromyography during a spontaneous breathing trial to predict extubation failure in preterm infants. Pediatric Res. 2022;92(4):1064–9. Epub 2022/05/07.10.1038/s41390-022-02085-w.

17. Kamlin CO, Davis PG, Argus B, Mills B, Morley CJ. A trial of spontaneous breathing to determine the readiness for extubation in very low birth weight infants: a prospective evaluation. Arch Dis Child Fetal Neonatal Ed. 2008;93(4):F305–6. https://doi.org/10.1136/adc.2007.129890.
18. Scott CB, Nickerson BG, Sargent CW, Platzker AC, Warburton D, Keens TG. Developmental pattern of maximal transdiaphragmatic pressure in infants during crying. Pediatric Res. 1983;17(9):707–9. https://doi.org/10.1203/00006450-198309000-00003.
19. Stark AR, Cohlan BA, Waggener TB, Frantz ID 3rd, Kosch PC. Regulation of end-expiratory lung volume during sleep in premature infants. J Appl Physiol. 1987;62(3):1117–23.
20. Decramer M. Hyperinflation and respiratory muscle interaction. Eur Respir J. 1997;10(4):934–41.
21. Braun NM, Arora NS, Rochester DF. Force-length relationship of the normal human diaphragm. J Appl Physiol Respir Environ Exerc Physiol. 1982;53(2):405–12.
22. Spiesshoefer J, Boentert M, Tuleta I, Giannoni A, Langer D, Kabitz HJ. Diaphragm involvement in heart failure: mere consequence of hypoperfusion or mediated by HF-related pro-inflammatory cytokine storms? Front Physiol. 2019;10:1335. Epub 2019/11/22.10.3389/fphys.2019.01335.
23. Dassios T, Arattu Thodika FMS, Nanjundappa M, Williams E, Bell AJ, Greenough A. Diaphragmatic ultrasound and patent ductus arteriosus in the newborn: a retrospective case series. Front Pediatr. 2023;11:1123939. Epub 2023/04/01.10.3389/fped.2023.1123939.
24. Maynard V, Bignall S, Kitchen S. Effect of positioning on respiratory synchrony in non-ventilated pre-term infants. Physiother Res Int. 2000;5(2):96–110.
25. Dimitriou G, Tsintoni A, Vervenioti A, Papakonstantinou D, Dassios T. Effect of prone and supine positioning on the diaphragmatic work of breathing in convalescent preterm infants. Pediatr Pulmonol. 2021;56(10):3258–64. Epub 2021/07/31.10.1002/ppul.25594.
26. Kassim Z, Greenough A, Rafferty GF. Effect of caffeine on respiratory muscle strength and lung function in prematurely born, ventilated infants. Eur J Pediatr. 2009;168(12):1491–5. https://doi.org/10.1007/s00431-009-0961-9.
27. Williams EE, Hunt KA, Jeyakara J, Subba-Rao R, Dassios T, Greenough A. Electrical activity of the diaphragm following a loading dose of caffeine citrate in ventilated preterm infants. Pediatr Res. 2019; https://doi.org/10.1038/s41390-019-0619-x.
28. Nava S, Gayan-Ramirez G, Rollier H, Bisschop A, Dom R, de Bock V, et al. Effects of acute steroid administration on ventilatory and peripheral muscles in rats. Am J Respir Crit Care Med. 1996;153(6 Pt 1):1888–96. https://doi.org/10.1164/ajrccm.153.6.8665051.

29. Dassios T, Kaltsogianni O, Dixon P, Greenough A. Effect of maturity and infection on the rate of relaxation of the respiratory muscles in ventilated, newborn infants. Acta Paediatr. 2018;107(4):587–92. https://doi.org/10.1111/apa.14188.
30. Song Y, Karisnan K, Noble PB, Berry CA, Lavin T, Moss TJ, et al. In utero LPS exposure impairs preterm diaphragm contractility. Am J Respir Cell Mol Biol. 2013;49(5):866–74. Epub 2013/06/26.10.1165/rcmb.2013-0107OC.
31. Stein H, Firestone K. Application of neurally adjusted ventilatory assist in neonates. Semin Fetal Neonatal Med. 2014;19(1):60–9. Epub 2013/11/19.10.1016/j.siny.2013.09.005
32. McKinney RL, Keszler M, Truog WE, Norberg M, Sindelar R, Wallstrom L, et al. Multicenter experience with neurally adjusted ventilatory assist in infants with severe bronchopulmonary dysplasia. Am J Perinatol. 2021;38(S 01):e162–e6. Epub 2020/03/26.10.1055/s-0040-1708559.
33. Kuitunen I, Rasanen K. Non-invasive neurally adjusted ventilatory assist (NIV-NAVA) reduces extubation failures in preterm neonates – a systematic review and meta-analysis. Acta Paediatr. 2024;113(9):2003–10. Epub 2024/05/04.10.1111/apa.17261.

Control of Respiration 10

The respiratory system is controlled by the respiratory centres in the pons and the medulla. These centres collect information from a network of receptors that are located in the central nervous system (central receptors), in the carotid and aortic bodies (peripheral receptors) or in the lung (stretch, irritant and others). The peripheral chemoreceptors in the carotid and aortic bodies respond rapidly to an increased partial pressure of carbon dioxide or abnormal pH and to a decreased partial pressure of oxygen. The ventilatory response to the stimulation of the pulmonary stretch receptors is a decrease in the respiratory frequency due to an increase in the expiratory time [1]. This phenomenon is called the *Hering-Breuer inflation reflex*.

The control of respiration is not fully developed in newborn infants and this phenomenon is more pronounced in prematurely-born infants who typically exhibit periodic breathing and long pauses of breathing, described as apnoea of prematurity [2]. When oxygen levels in the blood (partial pressure of arterial oxygen) decrease, the chemoreceptors in the carotid bodies respond by sending signals to the respiratory centre to increase the rate of breathing and the tidal volume. This phenomenon is called *hypoxic ventilatory response* and is blunted in prematurely born infants during their early postnatal life [3].

Infants of mothers who were smoking and/or misusing substances during pregnancy have dampened ventilatory responses to hypoxia and this dampening is more pronounced in the prone compared to the supine position [4]. This phenomenon, among others, might contribute to the pathophysiology on *sudden infant death syndrome* and explains the recommendation for discharged infants to sleep in the supine position once at home and not further monitored with saturation and hear rate measuring devices [5]. Apnoea in premature infants appears to be of mixed obstructive and central origin. The application of continuous positive airway pressure can reduce both obstructive and mixed apnoea, indicating that the obstructive element might be the predominant contributor in mixed events [6]. Other than an abnormal response to hypoxia, the pathophysiology of apnoea in preterm infants can also be explained by altered chemoreceptor responses to hypercapnia, as infants with apnoea exhibit a relatively flat slope on the carbon dioxide (CO_2) response curve [7].

Another potential mechanism might be that both preterm and term infants breathe closer to their CO_2 apnoeic threshold (the level of CO_2 below which breathing ceases) compared to adults, and this narrow range of CO_2 between apnoea and normal breathing might predispose these infants to apnoea and periodic breathing [8]. The adaptation of the prematurely-born infant to the extrauterine environment requires a complex adjustment to postnatal hypoxic and hypercapnic stimuli which render the respiratory control prone to instability [9].

Question

Question 1: The control of respiration:
 (a) Is modulated only by central receptors
 (b) Is independent of the level of the carbon dioxide
 (c) Is less developed in premature infants
 (d) Immature control in premature infants means they will have purely obstructive apnoea

References

1. West JB. Respiratory physiology : the essentials, vol. viii. 9th ed. Philadelphia: Wolters Kluwer Health/Lippincott Williams & Wilkins; 2012. 200 pp.
2. Poets CF, Samuels MP, Southall DP. Epidemiology and pathophysiology of apnoea of prematurity. Biol Neonate. 1994;65(3–4):211–9. Epub 1994/01/01.10.1159/000244055.
3. Debevec T, Pialoux V, Millet GP, Martin A, Mramor M, Osredkar D. Exercise overrides blunted hypoxic ventilatory response in prematurely born men. Front Physiol. 2019;10:437. Epub 2019/05/02.10.3389/fphys.2019.00437.
4. Rossor T, Ali K, Bhat R, Trenear R, Rafferty G, Greenough A. The effects of sleeping position, maternal smoking and substance misuse on the ventilatory response to hypoxia in the newborn period. Pediatr Res. 2018;84(3):411–8. Epub 2018/07/07.10.1038/s41390-018-0090-0.
5. Poets CF, von Bodman A. Placing preterm infants for sleep: first prone, then supine. Arch Dis Child Fetal Neonatal Ed. 2007;92(5):F331–2. Epub 2007/08/23.10.1136/adc.2006.113720.
6. Miller MJ, Carlo WA, Martin RJ. Continuous positive airway pressure selectively reduces obstructive apnea in preterm infants. J Pediatr. 1985;106(1):91–4. Epub 1985/01/01.10.1016/s0022-3476(85)80475-3.
7. Gerhardt T, Bancalari E. Apnea of prematurity: I. Lung function and regulation of breathing. Pediatrics. 1984;74(1):58–62. Epub 1984/07/01.
8. Khan A, Qurashi M, Kwiatkowski K, Cates D, Rigatto H. Measurement of the CO2 apneic threshold in newborn infants: possible relevance for periodic breathing and apnea. J Appl Physiol. 2005;98(4):1171–6. Epub 2005/03/18.10.1152/japplphysiol.00574.2003
9. Rigatto H. Control of ventilation in the newborn. Annu Rev Physiol. 1984;46:661–74. Epub 1984/01/01.10.1146/annurev.ph.46.030184.003305.

Waveforms   11

One clinical application where respiratory physiology meets bedside critical neonatal care are the ventilator waveforms which are graphs of the pressure, volume, flow and carbon dioxide dynamic signals displayed on the ventilator screen in real time [1].

Ventilator waveform monitoring is possible due to two major technological advancements:

1. Accurate and low dead space **sensors** that measure flow, pressure and carbon dioxide. These sensors need to be of a sufficiently *low dead space* so that they do not influence gas exchange. For example, a relatively large flow sensor in a very small infant will contribute a significant amount of dead space in the respiratory circuit and cause rebreathing of the exhaled gas from the previous breath and the presence of the sensor will inadvertently elevate the level of carbon dioxide in that infant. These sensors also need to be *proximal*, which means as close as possible to the infant in the respiratory circuit and typically between the endotracheal tube and the respiratory circuit so that the measurements are adequately accurate [2]. For example, in the previous generations of ventilators the pressure was measured in the main ventilator hardware (in the main ventilator box) so there was considerable loss of pressure from the ventilator into the long, heated and humidified respi-

ratory circuit. That the sensors ideally need to be as close as possible to the gas exchange part of the lungs can be demonstrated by the enhanced accuracy of the measurements which can be obtained with even more distal sampling, such as for example in the case of sampling carbon dioxide via a double lumen endotracheal tube which can sample at the level of the tracheal carina [3]. The sensors finally need to be *accurate* and *fast* enough, which means that they must have a high *sampling frequency* as the respiratory events are fast especially in preterm infants and of sufficient *sensitivity* as the actual delivered and measured parameters are also very small, for example a targeted tidal volume of 5 ml/kg in an infant of 500 gr is only 2.5 mls.
2. Ventilator **microprocessors** are also needed, which enable not only the processing of the displayed variables in real time but also the interaction of the measured variables with the ventilator software. This interaction is required for example in volume targeted ventilation, where the measured flow signal is integrated over time to derive the expired volume and this measured parameter is then used to deliver the next inflation at a lower or higher pressure so that the next tidal volume will be delivered as close as possible to the targeted value for the tidal volume.

It is important to remember that the mechanical ventilators measure some primary parameters and then some derived (secondary) parameters are calculated. The primary measured parameters are time, pressure and flow, while some ventilators can also measure and display the exhaled or tidal carbon dioxide. From these values, the volume is calculated via integration of the flow signal over time and some other composite indices can be calculated such as the compliance and the resistance of the respiratory system and the inspiratory time constant as discussed in the chapter on Respiratory Mechanics.

The ventilator waveforms can be displayed either as the measured parameter (pressure, flow, volume) over time or as a combination of these parameters of which the most commonly used are the pressure versus volume (PV curve) and the flow versus volume (FV curve) [1]. The North American literature also refers to the waveforms of pressure, flow and volume over time as scalars and the PV and FV curves as PV or FV loops.

Pressure Versus Time

The depiction of the pressure signal over time has the main advantage of schematically presenting the mean airway pressure (MAP). The MAP is the integration of the airway pressure over time as depicted in Fig. 11.1. The MAP in any form of mechanical ventilation is the most important determinant of oxygenation, second only to the actual fraction of the inspired oxygen which is being provided. The general rule is that the higher the MAP, the better the oxygenation. In conventional ventilation, if one aims to improve oxygenation, any intervention which will increase the area of the MAP in the graph (the area under the curve), will also improve oxygenation. Such interventions are the increase of the peak inflation pressure (PIP), the increase of the positive end expiratory pressure (PEEP) or the increase in the inspiratory time. In the previous generation of ventilators one could also alter the

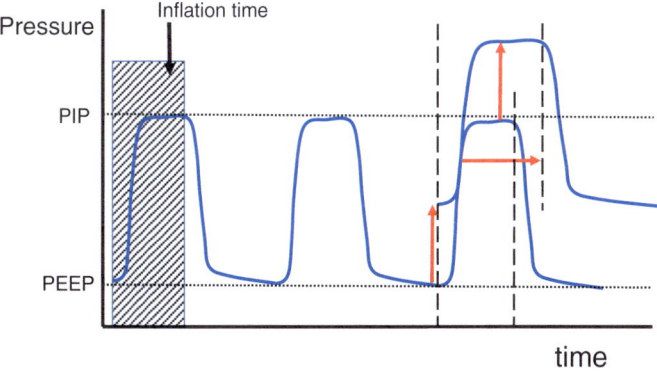

Fig. 11.1 The pressure versus time curve. The effect of increasing the peak inflation pressure (PIP), the positive end expiratory pressure (PEEP) and the inflation time on the mean airway pressure is highlighted with the arrows

inspiratory flow and this would increase the steepness of how the peak pressure is delivered, but this was reserved as an extreme measure as a very steep (abrupt/violent) provision of a high peak inflation pressure is thought to be injurious for the lung parenchyma and may cause air leak complications. The equivalent of a higher flow in the modern generation of ventilators is the rise time, which can also be adjusted with a similar effect and possibly with similar unintended actions. A short rise time is thought to be associated with improved oxygenation as it is associated with higher MAP with increased area under the pressure curve [4]. It is theoretically, however, also considered to be a more abrupt way to ventilate and could potentially cause some degree of lung injury. The rise time can also be altered with a view to achieve the intended inflation time, as a very long rise time might be insufficient for the pressure to reach the desired value within the prescribed inflation time, and thus a shorter/faster rise might be required.

Although beyond the scope of this chapter, it is also important to remember that all extremes in the delivered values come with specific anticipated side effects and should be avoided. For example a very high peak inflation pressure can damage the lung parenchyma and cause pneumothorax, or overexpansion and inefficient gas exchange, but a very small peak inflation pressure can be insufficient for gas exchange and be associated with atelectasis. Similarly, a very short inflation time might be insufficient to reach the desired pressure and insufficient for adequate oxygenation, or contrarily a very long inflation time might be injurious to the pulmonary parenchyma, cause asynchrony and not allow for sufficient time for emptying of the lung at a set ventilation rate which might be required for effective carbon dioxide clearance.

The pressure versus time waveform can help the clinician to visually differentiate the breaths in synchronised ventilation, as the triggered breaths are usually presented on the ventilator screen in a different colour than the backup rate, while when on synchronised intermittent mandatory ventilation, where the triggered breaths are not supported, the pressure waveform of the triggered breaths (the spontaneous breaths) will be of significantly smaller magnitude compared to the backup ones (Fig. 11.2). When an infant is spontaneously breathing on invasive ventilation, there is some value in the rate of decline of the pressure signal over time,

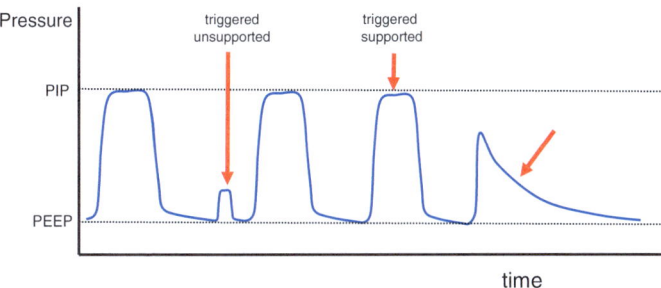

Fig. 11.2 Triggering on the pressure versus time curve. Triggered unsupported breaths can either be supported or unsupported depending on the mode of ventilation. A delay of the pressure signal to return to baseline (diagonal arrow) represents respiratory muscle fatigue

with a slow rate of decline (i.e. a longer time is required for the spontaneous pressure signal to return to zero) signalling impaired muscle function and an increased risk of muscle fatigue, and this information has potential clinical applicability in predicting the success of an extubation attempt in premature infants [5].

Flow Versus Time

The flow versus time waveform typically consists of two phases. By convention, the positive part of the waveform corresponds to the inspiration (or inflation) and the negative to the expiration (or deflation). The terms in parentheses are equivalent in the literature, the terms inflation/deflation, however, are preferentially used for ventilated subjects, rather than inspiration/expiration which imply some spontaneous effort from a subject who is not being artificially supported in their breathing effort [6]. When on synchronised ventilation where the triggered breaths are fully supported similarly to the back-up inflations (called SIPPV or PTV or Assist/Control), if the generated spontaneous flow of the infant—the respiratory effort—is lower than the trigger threshold, then the flow sensor will fail to detect the breath and the breath will not be supported (failure to trigger). The solution to this problem is to lower the trigger sensitivity, which is an index that depends on the

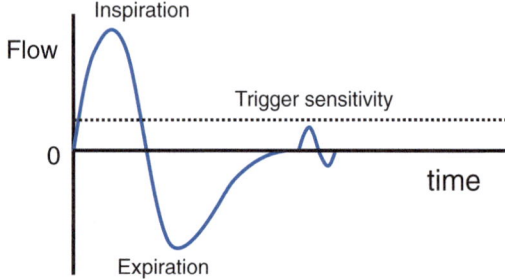

Fig. 11.3 Failure to trigger. The flow generated during a spontaneous breathing effort is below the detection threshold (trigger sensitivity)

infant. It is useful to know that the average peak inspiratory flow rate in a spontaneously breathing term infant is approximately 2.9 L/min [7], so very low trigger thresholds are not always required especially in larger infants. It follows that a normal term and unsedated infant might be comfortable with a trigger sensitivity of 1 L/min but a very small infant might require a significantly lower trigger threshold in the area of 0.4 L/min (Fig. 11.3). Of note, very low trigger rates are also to be avoided due to the phenomenon of *auto-triggering*, where the oscillation of condensed air forming water droplets in the respiratory circuit might generate small flows which can be erroneously recognised by the flow sensor as breathing efforts, increasing the total rate of ventilation to very high numbers sometimes exceeding 100 per minute. The problem at such high rates is that ventilation becomes inefficient as complete respiratory cycles with an adequate inspiratory plateau for oxygenation and full deflation for efficient carbon dioxide clearance are difficult to achieve in such a short time per breath/inflation [8]. The solution to this problem is the emptying of the droplets from the circuit into the water trap and the adjustment if required, of the humidity of the circuit.

As previously discussed, the process of emptying of gas from the lungs during deflation is inherently more time-consuming and mechanically more cumbersome than the filling of the same compartment with gas during inflation. When there is some element of airway obstruction, such as during a disease characterised by

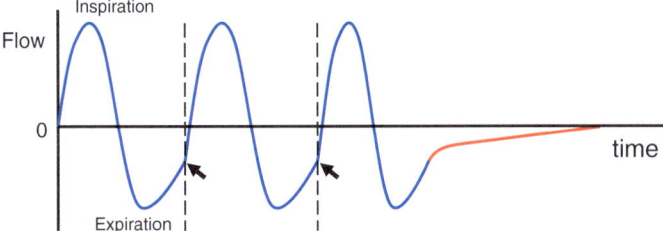

Fig. 11.4 Flow versus time curve. Insufficient time for deflation due to prolonged expiration. The flow signal fails to return to zero after deflation and before the next inflation

increased respiratory resistance, such as bronchiolitis in infancy or severe BPD in the newborn population, the expiratory part of the flow versus time waveform will appear prolonged and will take longer to reach zero compared to the inspiration/inflation phase (Fig. 11.4). This can also happen when there is a very long and narrow endotracheal tube (for example a size 2.0 mm in the tiniest preterms) or when acute mechanical complications increase the airway resistance such as with increased secretions, mucous plugs, old blood in the endotracheal tube or if the tube is kinked, and the resistance appears suddenly increased. At extremes of such cases of high expiratory resistance, it is possible that the expiratory flow does not return to zero before the beginning of the next inflation, which might lead to accumulation of residual air in the lung due to incomplete emptying. In these cases and if the endotracheal tube is particularly tightly fitting in the trachea, this might cause progressive retaining of gas in the lungs and subsequently lead to pneumothorax or other air leak complications. The solution to this problem is to use endotracheal suction to remove the materials which coat the tube and the airways as much as possible, and to decrease the rate of ventilation so that sufficient time is allowed for full emptying of the lungs in the presence of increased resistance. This is why typically infants with severe BPD or bronchiolitis are ventilated with low rates and long inspiratory and expiratory times.

Turbulent flow secondary to increased humidity and condensed water in the circuit will also manifest in the flow versus time waveform as small flow oscillations which might impact the efficiency of the delivered ventilation or cause auto-triggering and this can be mitigated by emptying of the circuit or decreasing the humidity.

Volume Versus Time

In the volume versus time waveform, one can visually assess the provided tidal volume during the respiratory cycle. The inspiratory tidal volume is the first rising part of the waveform and the expiratory tidal volume the second declining part, which tends to return back towards zero (Fig. 11.5). It is useful to remember that the flow sensor usually has two pre-heated fine wires, one closer to the infant and the other one more distal from the infant, and this way the ventilator can discriminate if the flow is incoming or outgoing: If the proximal wire is the first to cool down with exhaled air, then this is recognised by the ventilator as an expiration, and if the distal wire cools down before the one closer to the infant, this flow is recognised as an inflation from the ventilator.

The difference between the inspiratory and the expiratory part of the measured tidal volume is used to calculate the percentage of leak. Most modern ventilators can adjust volume targeted ventilation to high leak values but irrespective of whether ventilation is provided in volume targeted or pressure controlled mode, if there is a very high leak exceeding 40% or 50% it is worth remembering that the provision of ventilation is not as efficient as it would

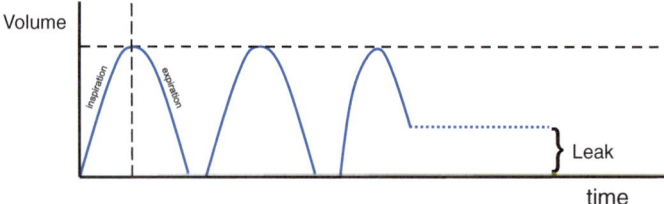

Fig. 11.5 Volume versus time curve. Inspiration (or inflation) and expiration (or deflation) and the endotracheal tube leak are depicted

be with a smaller leak. In these circumstances it is possible that a wider endotracheal tube is required (Fig. 11.5) [9].

Significant acute changes in the delivered volume while on pressure-controlled ventilation should raise the suspicion of an acute mechanical complication. For example, if on a given peak inflation pressure the measured volume was consistently 10 ml and suddenly drops down to around 5 ml there is a possibility that the tube has slipped into the right main bronchus and only the right lung is now being ventilated. Confirmation of the tube position and adjustment of the endotracheal tube length should be sufficient to solve this problem.

An acute decrease in the delivered tidal volume while on pressure-controlled ventilation, or an acute increase in the required pressure to achieve the targeted tidal volume on volume targeted ventilation, might also be the result of acute obstruction of the tube such as with a kinked tube, or an endotracheal tube partially obstructed by an inline suction catheter. When this happens during volume targeted ventilation, the flow sensor will progressively read lower exhaled volumes, and the inflation pressure will be increased so that the targeted tidal volume is achieved. When this pressure reaches the alarm threshold of the maximum peak inflation pressure or PIP_{max} (usually set at 5 cm H_2O above the required inflation pressure during normal circumstances) then the alarm will also go off. Careful inspection or suction might completely alleviate this problem (unkinking or suctioning of the tube) and the pressure requirements will then return back to the previous expected values.

Pressure Versus Volume

In all intensive care medicine, historically the pressure-volume (PV) curve is thought to be the one that gives the largest amount of information on the mechanics and the overall respiratory status of a ventilated individual at a given point of time. As described in an earlier chapter, the compliance of the respiratory system can be depicted in the PV curve by the overall slope of the curve in relation to the y axis. A relatively upright slope signifies normal lung

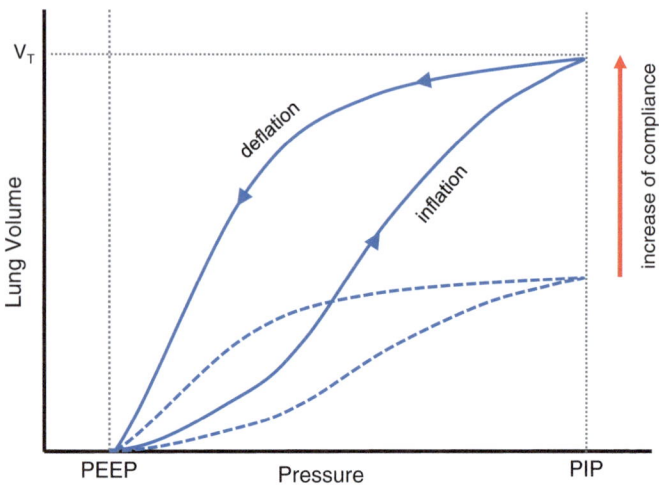

Fig. 11.6 The pressure-volume (PV) curve of a ventilated infant. The positive end expiratory pressure (PEEP), peak inflation pressure (PIP) and tidal volume (V_T) are depicted. The PV curve of a preterm infant with respiratory distress syndrome will be initially depressed and will "move upwards" after administration of surfactant and an increase in compliance

mechanics and good compliance, while a depressed curve signifies poor compliance and stiff lungs (Fig. 11.6). In neonatal medicine, very commonly a preterm infant with respiratory distress syndrome (RDS) will start with a very depressed, short and elongated curve and with the administration of surfactant this curve will move upwards reflecting an improvement in compliance [10].

Since the goal of ventilation is gas exchange which is achieved by volume change (moving a volume of air in and out of the lungs), if the PV curve demonstrates an area where significant pressure changes do not correspond to volume change, this area of the graph then depicts inefficient ventilation with unnecessary high pressures which do not achieve any volume change. This graph resembles the beak of a bird, and the phenomenon is then referred to as *beaking of the PV curve*. The solution to this phenomenon is to "cut off the beak" by decreasing the inflation pres-

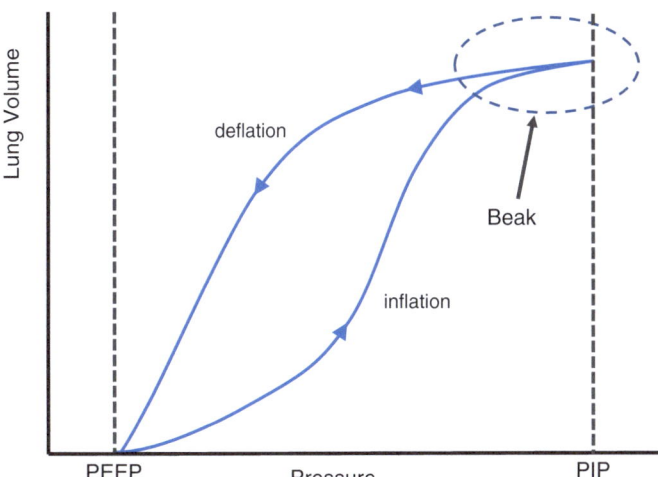

Fig. 11.7 Beaking of the curve. Excessive provision of pressure which does not translate to significant volume change flattens the pressure volume curve at end inflation in a way that visually resembles a beak of a bird. PEEP positive end expiratory pressure, PIP peak inflation pressure

sure to the minimum value which still produces some reasonable volume change (Fig. 11.7).

What we observe on the ventilator screen is the PV curve while that infant is ventilated with the parameters that we have set at that point of time, for example with a peak pressure of 20 and a positive end expiratory pressure of 5 cm H_2O. There are, however, parts of the PV relationship which we cannot see as they correspond to pressure and volume values which we are not delivering at that point. If we could plot the pressure and volume for all possible provided values starting from a pressure of 0 and going up to excessively high pressures we would get a fuller understanding of the mechanical properties of the lung at that specific time (Fig. 11.8).

One can then observe that there is a zone of optimal recruitment where the rate of change of volume is optimal (high) in relation to the inflation pressures and two extreme zones at either side of the optimal/open lung zone where ventilation is less efficient,

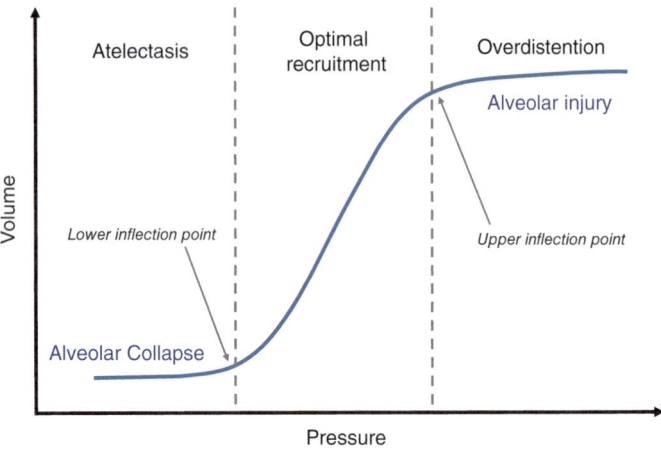

Fig. 11.8 Setting the peak inflation and positive end expiratory pressures. Pressure—Volume curve demonstrating that ventilation is most efficient and volume change is maximised for a given pressure change, in an intermediate zone avoiding both extremes of atelectasis or overinflation

i.e. despite giving a lot of pressure we achieve little or no volume change because the lung is either collapsed and atelectatic or overinflated. The practical utility of this graph is that the two points of inflection can guide us on what pressures we need to use to ventilate in this optimal zone. The first inflection point (lower inflection point) can be used to set the positive end expiratory pressure as this is the minimum pressure required to keep the lung open, and the second inflection point (upper inflection point) can be used to set the peak inflation pressure as this is the maximum pressure that we can use, which effectively moves air in and out and above this value we are only injuring the lung without achieving any volume change or gas exchange [10].

Another morphological part of interest in the PV curve is the difference between the inflation and deflation limb of the curve which is called *hysteresis*. The phenomenon that the inflation and the deflation limbs are not identical can be explained by the fact that the lung is not a perfect elastic material and exhibits slightly different mechanical properties when stretched during expansion

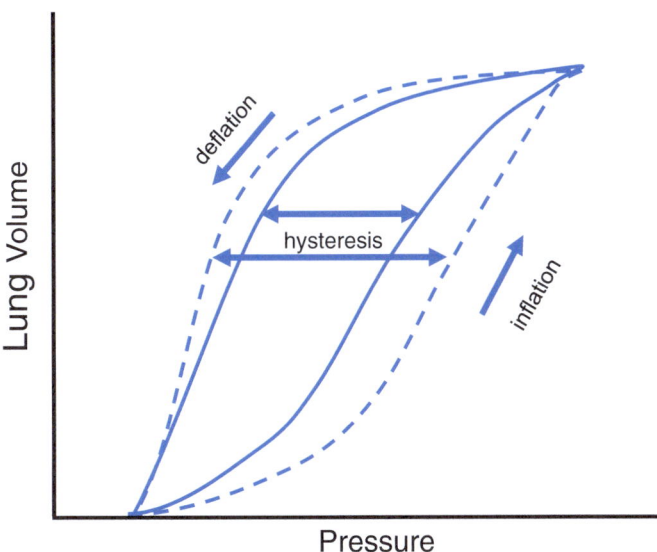

Fig. 11.9 Hysteresis. The difference between the inflation and deflation limb of the curve is an index of airway resistance

compared to when it returns to its resting state. The hysteresis is an index of increased resistance and an increased distance between the two parts of the curve signifies increased resistance of the airways (Fig. 11.9).

Similarly to what we described above in the flow and volume versus time waveforms, acute changes which grossly affect the delivery of the intended tidal volume will also manifest with changes in the PV curve. It might be worth remembering that in pressure-controlled ventilation, a complication such as right main bronchus intubation, will manifest with halving of the volume while the pressure will remain constant, while contrarily in volume-targeted ventilation, the same complication will manifest with a gradual increase in the provided pressure while the volume will remain fairly constant as it is being targeted to remain stable (Fig. 11.10).

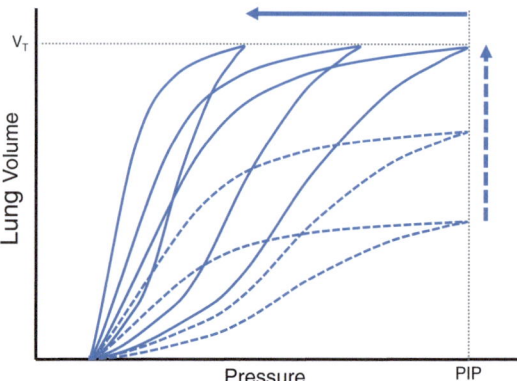

Fig. 11.10 The Pressure-Volume curve in pressure-limited and volume-targeted ventilation. Improving compliance on volume-targeted ventilation (solid arrow) is depicted by decreasing pressures to achieve the same tidal volume. Improving compliance in pressure-limited ventilation (dashed arrow) is depicted by increasing volumes while the same pressure is applied. V_T tidal volume, PIP peak inflation pressure

Flow Volume Curve

The flow volume (FV) curve typically depicts the flow on the y axis and the volume on the x axis and by convention the inspiratory part of the inflation is depicted as negative and the expiratory part as positive. In health, the inspiratory part of the flow volume curve is usually circular denoting normal laminar flow through the airways. In conditions of increased resistance the expiratory part of the curve will demonstrate a steep part (Fig. 11.11) which corresponds to the limitation of expired flow though stenotic or narrow airways. This phenomenon in extreme resembles a ski slope and is thus referred to as a "ski slope" FV curve.

If the volume at end expiration is not yet zero before the beginning of the next inflation this describes a mechanism for potential air trapping as also described in the flow versus time waveform, and if the volume after expiration returns to a value which is smaller compared to the volume measured during inspiration/inflation this difference can be used to visually quantify and depict air leak (Fig. 11.12). Another potential application of the flow vol-

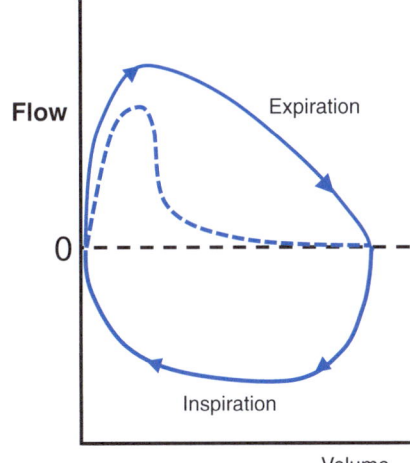

Fig. 11.11 Flow volume curve. The expiratory part of the curve becomes steeper in conditions of increased respiratory resistance (dashed line)

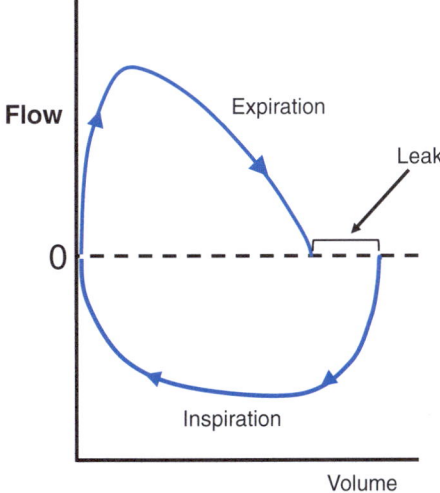

Fig. 11.12 Endotracheal tube leak in a flow volume curve. The expiratory limb of the curve fails to return to the same volume point from where the inspiratory limb started

Fig. 11.13 Titrating continuous positive airway pressure (CPAP). Effect of different levels of CPAP on the shape of the flow volume curve in conditions of expiratory obstruction, such as tracheomalacia. Incremental pressure changes (steps a and then b) demonstrate that the airways gradually open up more and remain patent during expiration

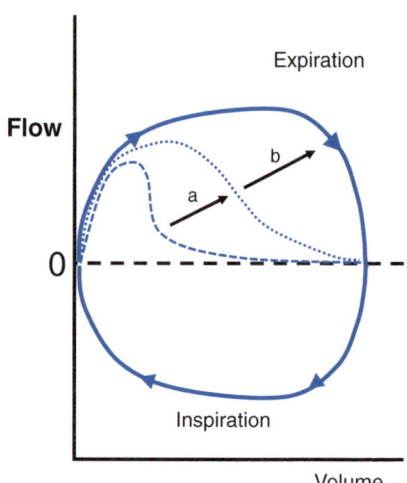

ume curve in infants with collapsible airways during expiration (such as for example in laryngo-bronchomalacia or tracheomalacia) is the titration of the positive end expiratory pressure required to keep the airways open at end expiration with a view to maximise lung volumes and oxygenation. As depicted in Fig. 11.13, a gradual increase in the positive end expiratory pressure during non-invasive respiratory support will result in a gradual normalisation of the expiratory limb of the FV curve signifying better emptying of air at expiration and avoidance of airway collapse.

Carbon Dioxide Versus Time

Although historically there has been some significant scepticism regarding the accuracy and utility of tidal capnography in neonatal respiratory care, recent developments of low dead space and fast and accurate side stream and mainstream capnographs have made the use of such technology more reliable and acceptable in neonatal intensive care [11].

Exhaled tidal carbon dioxide (CO_2) can be depicted over time on the ventilator screen or on monitoring screens. The maximum

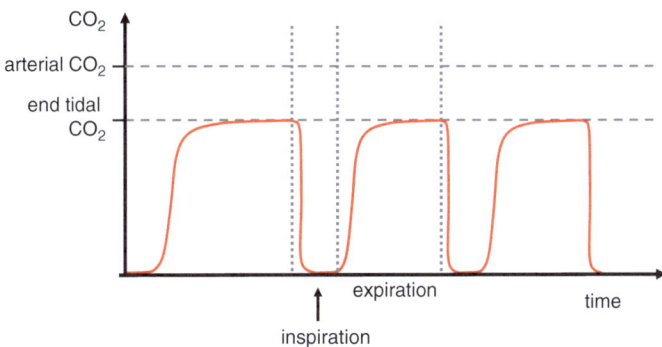

Fig. 11.14 Carbon dioxide versus time. A capnogram of a ventilated infant with normal lungs and a plateau phase where exhaled carbon dioxide (CO_2) is fairly constant. The arterial CO_2 is always a larger value compared to the end-tidal CO_2

value of the CO_2 corresponds to the end-tidal CO_2 and is always lower than the arterial CO_2 as it is diluted with gas from non gas-exchanging areas of the lung and the apparatus which also does not produce CO_2 [12]. Of note, and although not strictly related to this chapter, it is also worth remembering that transcutaneous CO_2 is relatively higher than both arterial and end tidal CO_2 as it describes the CO_2 state at the level of the tissues where gas exchange is already happening or has already happened, while arterial CO_2 is the CO_2 of the arterial gas and by definition CO_2 has not been yet collected from the peripheral tissues into the venous blood (Fig. 11.14). In ventilated preterm infants, the correlation between arterial and end tidal CO_2 is acceptable but the two values would not be expected to be the same. On the contrary, although the values are well correlated, there is at the same time an expected and fixed difference, which has been reported at a mean value of 0.54 kPa (or 4 mmHg) between the arterial and the end tidal CO_2 values [11].

Irrespective of the appearance of the waveform, the difference between the arterial and the end tidal CO_2 holds some clinical information. In the chapter on perfusion we explained how this difference is an indication of the total respiratory dead space (or physiological dead space), as the larger the dead space the more diluted the exhaled CO_2 would be compared to the arterial CO_2, and if this difference is known, one can work backwards to estimate the actual total dead space. This total or *physiological dead space* includes both the anatomical and the alveolar compartments. In the same individual, dynamic changes in the difference between the end tidal and the arterial CO_2 can be attributed to evolving changes in the alveolar dead space which is an index of lung disease severity. For example, we previously discussed that ventilated infants with persistent pulmonary hypertension of the newborn and recirculation of CO_2-rich venous blood into the systemic circulation will clinically present with an increased arterial to end tidal CO_2 gradient of approximately 10 mmHg, which will decrease to nearly 3 mmHg following resolution of the disease [13].

The mere presence of a CO_2 waveform is a major safety indicator and source of reassurance as it guarantees that the endotracheal tube is correctly positioned in the trachea and exhaled CO_2 is detected by the sensor which is placed between the endotracheal tube and the ventilator tubing [14]. It is also worth remembering that exhaled CO_2 might not be detected in conditions of cardiorespiratory arrest even if the endotracheal tube is correctly placed in the trachea, as there is no metabolic production of CO_2 under these circumstances. This is useful knowledge in resuscitation scenarios, where removal of an endotracheal tube because of an absent CO_2 trace is not always the correct approach when there is no heart rate. It also follows than a sudden loss of the tidal CO_2 waveform might indicate that the tube has been dislodged or totally obstructed (Fig. 11.15).

When evaluating the appearance and morphology of a tidal capnogram over time, this graph will depict CO_2 only during exhalation or expiration and the inspiratory phase will generally correspond to an absence of CO_2, other than a brief rapid decrease of the CO_2 at the beginning of inspiration which is "leftover" CO_2 in the upper airways from the previous exhalation (Fig. 11.14).

Similar to the volume capnogram which we discussed in an earlier chapter, the *time capnogram* can also be divided into three phases (Fig. 11.16). Phase I is the first part of expiration during which gas is expired from the large conducting airways, this phase corresponds to little or no expired carbon dioxide. Phase II is the

Carbon Dioxide Versus Time

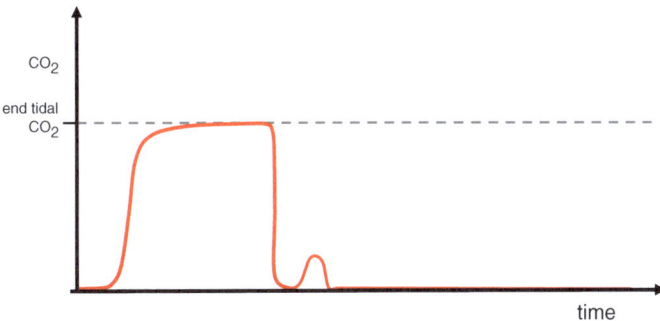

Fig. 11.15 Loss of CO_2 signal. The endotracheal tube might have dislodged or might be totally obstructed

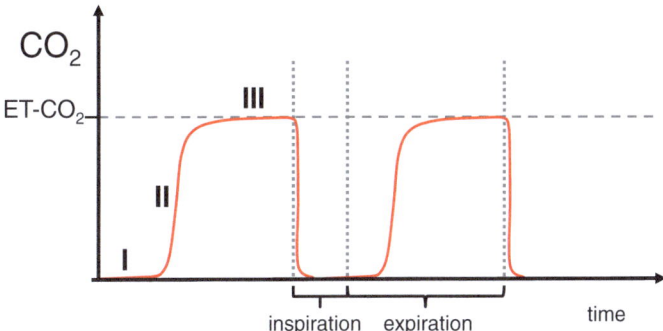

Fig. 11.16 The phases of a time capnogram. ET-CO_2: end tidal carbon dioxide

second part where there is mixing of gas from the large and small airways, thus exhaled carbon dioxide levels rapidly rise. Phase III is the final part which contains pure alveolar gas; it is at the end of this IIIrd phase where the end-tidal carbon dioxide (EtCO$_2$) value is determined. A steep (upright) phase III occurs if there is ventilation inhomogeneity, as different lung units with differing time constants empty unevenly during expiration [15]. The alveolar plateau might be shorter in preterm infants with high breathing rates or absent in spontaneous breaths during synchronised ventilation as expiration might finish before complete emptying of the

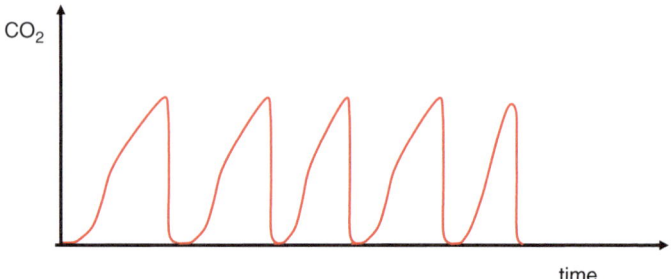

Fig. 11.17 "**Shark fin**" **capnogram.** Upward slant in phase III corresponding to increased expiratory resistance

alveolar gas: in these cases, the end tidal CO_2 will underestimate the "true" alveolar CO_2.

In conditions of increased expiratory resistance, there can be observed a consistent upward slant in phase III, which has been described as a "shark fin" and pathophysiologically corresponds to ventilation inhomogeneity and increased expiratory resistance such as can happen in bronchopulmonary dysplasia (Fig. 11.17).

Rebreathing

A rising baseline with the end tidal CO_2 value not returning to zero at inspiration might describe increased re-breathing of CO_2 (breathing of expired gas) which occurs when the tidal volume is set too low or when there is increased apparatus dead space, leading to ineffective removal of CO_2 from the respiratory circuit. The end tidal CO_2 value will be depicted as gradually rising with consecutive breathing cycles (Fig. 11.18).

Breathing out of synchrony with the ventilator, could cause a "dent" pattern on the waveform as the infant tries to inspire during expiration causing a small decrease of the CO_2 in the waveform (Fig. 11.19).

Decreasing end tidal CO_2 levels with increasing breathing frequency are observed in hyperventilation and, contrarily, an increasing end tidal CO_2 with slowing down of the breathing frequency are seen during hypoventilation (Figs. 11.20 and 11.21).

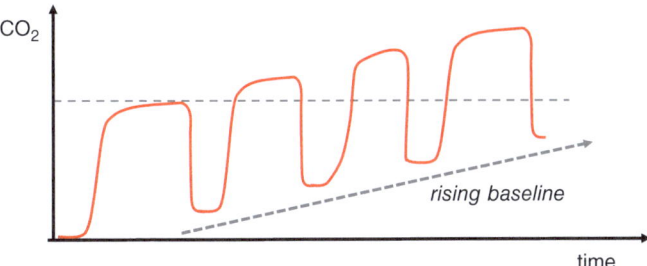

Fig. 11.18 Rebreathing capnogram. A rising baseline with the exhaled CO_2 not returning to zero at inspiration

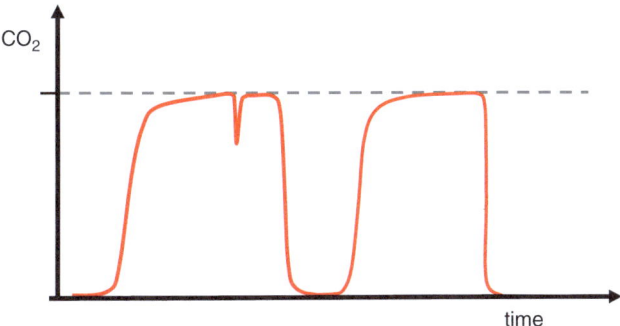

Fig. 11.19 Dented capnogram. A spontaneous breath during deflation could cause a dent appearing mid-plateau. This pattern signifies breathing out of synchrony with the ventilator

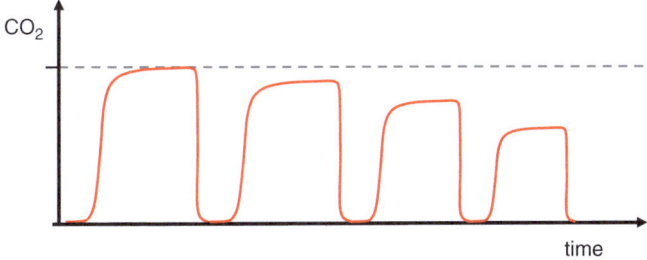

Fig. 11.20 Hyperventilation on time capnography. Decreasing end tidal CO_2 levels and a high breathing frequency are depicted

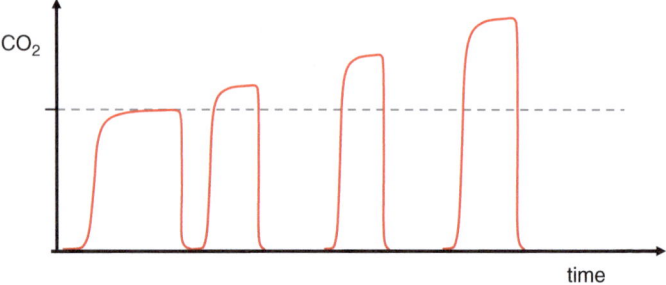

Fig. 11.21 Hypoventilation on time capnography. Increasing end tidal CO_2 levels and a low breathing frequency are depicted

Carbon Dioxide Versus Volume

Other than the common depiction of the changing CO_2 value against time as described above, some ventilators and respiratory function monitors also offer the option to present in real time graphs of plotted expired CO_2 against the corresponding expired tidal volume during a single breath (Fig. 11.22). This graph is called a *volumetric capnogram* [16]. Similarly to the time capnogram, expiration on a volumetric capnogram can also be divided into three phases: phase I which corresponds to the initial part of expiration when only the large non-gas exchanging conducting airways are represented, phase II which describes the mixed emptying of conducting airways and alveolar units and phase III which represents the concurrent emptying of gas from the alveoli and gas-exchanging distal airways [17]. Volumetric capnography can be used for offline analysis and the calculation of anatomical and alveolar dead space in ventilated infants [18]. The slopes of phases II (S_{II}) and III (S_{III}) also hold some important clinical information, as the slope can act as an index of ventilation inhomogeneity. Steeper slopes result from a larger degree of ventilation inhomogeneity as distal lung units empty unevenly creating a steeper upwards curve during phase III [17].

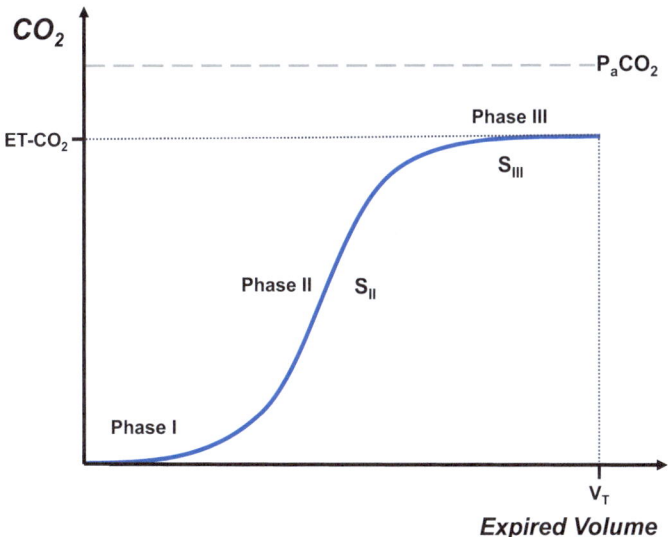

Fig. 11.22 Volumetric capnography. The three phases of a volumetric capnogram and the points corresponding to end tidal CO_2 (ET-CO_2), arterial CO_2 and expiratory tidal volume. The slopes of phases II (S_{II}) and III (S_{III}) are also depicted

The slope of phase III (S_{III}) has been positively related to indices of lung disease severity such as the duration of ventilation and required F_iO_2 to maintain normoxia in ventilated term and preterm infants and can potentially act as an index of disease severity in these newborns with preterm infants exhibiting steeper slopes compared to term ones [19] (Fig. 11.23). It has also been shown that in infants with bronchopulmonary dysplasia (BPD) the S_{III} was steeper compared to preterm infants without BPD and could discriminate these two categories significantly better than other indices of ventilation inhomogeneity such as the lung clearance index [20].

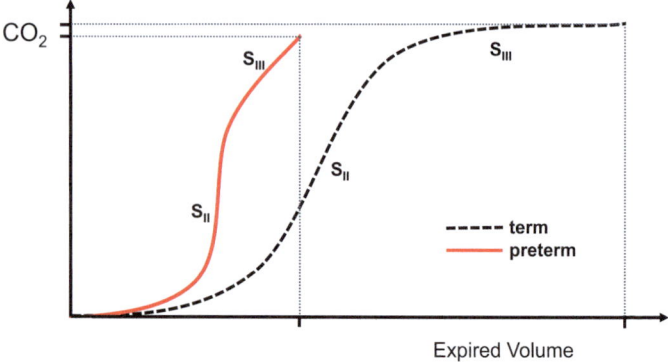

Fig. 11.23 Slopes of volumetric capnograms. Schematic representation of a typical volumetric capnogram in preterm and term infants, with preterm infants exhibiting steeper phases II and III compared to term infants

Questions

Question 1: If a flow sensor is faulty:
  (a) The volume waveform can still be displayed
  (b) Volume targeted ventilation can be given reliably
  (c) The pressure volume curve is accurate
  (d) Recalibration of the sensor can potentially resolve the issue

Question 2: The mean airway pressure
  (a) Increases with decreasing the positive end expiratory pressure because the pressure differential increases
  (b) Increases with increasing the peak inflation pressure
  (c) Is not affected by changes in the inspiratory time
  (d) Does not affect oxygenation

Question 3: In the pressure versus time waveform:
  (a) During synchronised intermittent mandatory ventilation all breaths/inflations will look the same
  (b) During synchronised intermittent positive pressure ventilation (or assist control or patient triggered ventilation) the triggered breaths will look smaller in magnitude compared to the backup inflations.

(c) It can be assessed whether the set pressure is achieved
(d) One cannot differentiate between a high and a low rise time.

Question 4: In the flow versus time waveform:
(a) Failure to trigger cannot be evaluated
(b) If there is increased resistance, the expiratory phase (deflation) will appear as having a shorter duration than the inspiration (inflation)
(c) Expiration should return to zero before the beginning of the next inspiration
(d) Turbulent flow will manifest with a long inflation time

Question 5: In the volume versus time waveform:
(a) The leak is visible as a failure of the declining part of the waveform to return to zero
(b) In volume-targeted ventilation the volume will fluctuate from breath to breath
(c) In pressure-controlled ventilation slipping of the endotracheal tube in the right main bronchus will manifest with doubling of the delivered volume
(d) A leak of 10% precludes the use of volume-targeted ventilation

Question 6: In the pressure-volume curve:
(a) Beaking of the curve refers to the condition where there is little pressure change while excessive volumes are given
(b) If the inspiratory and expiratory limbs appear very far apart this signifies increased airway resistance
(c) The upper inflection point is useful for setting the positive end expiratory pressure
(d) The pressure will remain constant if the compliance changes and we are ventilating in a volume-targeted mode.

Question 7: The following can be assessed in a flow volume curve:
(a) The compliance of the respiratory system
(b) Increased pressure requirements
(c) Increased expiratory resistance
(d) Whether the inflation is triggered or is a backup-one during invasive ventilation

Question 8: In the waveform of carbon dioxide against time:
(a) Sudden loss of signal means that there is ventilation inhomogeneity
(b) Cardiorespiratory arrest will not affect the waveform as long as the tube is in
(c) Has a steeper rise in bronchopulmonary dysplasia like a shark fin
(d) A rising baseline signifies hypoventilation

References

1. Donn SM, Mammel MC. Neonatal pulmonary graphics : a clinical pocket atlas. New York: Springer; 2015. xv, 175 p.
2. Schmalisch G. Current methodological and technical limitations of time and volumetric capnography in newborns. Biomed Eng Online. 2016;15(1):104. https://doi.org/10.1186/s12938-016-0228-4.
3. Kugelman A, Zeiger-Aginsky D, Bader D, Shoris I, Riskin A. A novel method of distal end-tidal CO2 capnography in intubated infants: comparison with arterial CO2 and with proximal mainstream end-tidal CO2. Pediatrics. 2008;122(6):e1219–24. Epub 2008/11/26. https://doi.org/10.1542/peds.2008-1300.
4. Chong D, Kayser S, Szakmar E, Morley CJ, Belteki G. Effect of pressure rise time on ventilator parameters and gas exchange during neonatal ventilation. Pediatr Pulmonol. 2020;55(5):1131–8. Epub 2020/03/10. https://doi.org/10.1002/ppul.24724.
5. Dassios T, Kaltsogianni O, Greenough A. Relaxation rate of the respiratory muscles and prediction of extubation outcome in prematurely born infants. Neonatology. 2017;112(3):251–7. Epub 2017/07/14. https://doi.org/10.1159/000477233.
6. Morley CJ, Keszler M. Ventilators do not breathe. Arch Dis Child Fetal Neonatal Ed. 2012;97(6):F392–4. Epub 2012/10/20. https://doi.org/10.1136/fetalneonatal-2012-302137.
7. Swyer PR, Reiman RC, Wright JJ. Ventilation and ventilatory mechanics in the newborn: methods and results in 15 resting infants. J Pediatr. 1960;56:612–22. Epub 1960/05/01. https://doi.org/10.1016/s0022-3476(60)80334-4.
8. Chakkarapani AA, Adappa R, Mohammad Ali SK, Gupta S, Soni NB, Chicoine L, et al. "Current concepts of mechanical ventilation in neonates" – Part 1: basics. Int J Pediatr Adolesc Med. 2020;7(1):13–8. Epub 2020/05/07. https://doi.org/10.1016/j.ijpam.2020.03.003.

9. Keszler M, Abubakar MK. Volume-targeted ventilation. Semin Perinatol. 2024;48(2):151886. Epub 2024/03/30. https://doi.org/10.1016/j.semperi.2024.151886.
10. Waugh JB, Harwood RJ. Rapid interpretation of ventilator waveforms. 3rd ed. Burlington, MA: Jones & Bartlett Learning; 2023. xvii, 163 p.
11. Williams E, Dassios T, Greenough A. Assessment of sidestream end-tidal capnography in ventilated infants on the neonatal unit. Pediatr Pulmonol. 2020;55(6):1468–73. Epub 2020/03/19. https://doi.org/10.1002/ppul.24738.
12. Williams E, Dassios T, Greenough A. Carbon dioxide monitoring in the newborn infant. Pediatr Pulmonol. 2021;56(10):3148–56. Epub 2021/08/09. https://doi.org/10.1002/ppul.25605.
13. Williams EE, Bednarczuk N, Nanjundappa M, Greenough A, Dassios T. Monitoring persistent pulmonary hypertension of the newborn using the arterial to end tidal carbon dioxide gradient. J Clin Monit Comput. 2024;38(2):463–7. Epub 2023/12/27. https://doi.org/10.1007/s10877-023-01105-2.
14. Williams EE, Dassios T, Harris C, Greenough A. Capnography waveforms: basic interpretation in neonatal intensive care. Front Pediatr. 2024;12:1396846. Epub 2024/04/19. https://doi.org/10.3389/fped.2024.1396846.
15. Williams EE, Dassios T, Hunt KA, Greenough A. Volumetric capnography pre- and post-surfactant during initial resuscitation of premature infants. Pediatr Res. 2022;91(6):1551–6. Epub 2021/05/24. https://doi.org/10.1038/s41390-021-01578-4.
16. Fowler WS. Lung function studies; the respiratory dead space. Am J Phys. 1948;154(3):405–16. https://doi.org/10.1152/ajplegacy.1948.154.3.405.
17. Fletcher R. Airway deadspace, end-tidal CO2, and Christian Bohr. Acta Anaesthesiol Scand. 1984;28(4):408–11. https://doi.org/10.1111/j.1399-6576.1984.tb02088.x.
18. Dassios T, Dixon P, Hickey A, Fouzas S, Greenough A. Physiological and anatomical dead space in mechanically ventilated newborn infants. Pediatr Pulmonol. 2018;53(1):57–63. https://doi.org/10.1002/ppul.23918.
19. Dassios T, Dixon P, Williams E, Greenough A. Volumetric capnography slopes in ventilated term and preterm infants. Physiol Meas. 2020;41(5):055001. Epub 2020/04/17. https://doi.org/10.1088/1361-6579/ab89c7.
20. Fouzas S, Hacki C, Latzin P, Proietti E, Schulzke S, Frey U, et al. Volumetric capnography in infants with bronchopulmonary dysplasia. J Pediatr. 2014;164(2):283–8 e1–3. https://doi.org/10.1016/j.jpeds.2013.09.034.

The Neonatal Respiratory System at Critical Extremes

Periviable Gestation

Historically it has been believed that a prerequisite for neonatal survival would be some—even minimal—development of the gas exchanging interface which consists of the alveolar structures, and thus no survival should be expected before 24 or 23 weeks of gestation. This concept has been challenged, however, by the lowering of the threshold of survival to 22 weeks of gestation [1] while some centres have reported survival even at 21 weeks [2]. This improved survival of extremely preterm infants, who up to recently were thought to be pre-viable, can be partly explained by advancements in the respiratory and overall management of these infants with proactive administration of antenatal steroids, preferential use of non-invasive respiratory support from birth, early surfactant and caffeine administration and minimising the duration of invasive mechanical ventilation when this is possible [3]. Although it is generally accepted that invasive ventilation is injurious to the developing preterm lung and that non-invasive support can be lung-protective [4, 5] most infants born before 24 weeks realistically will be intubated and ventilated at birth [6]. It is interesting to add that they might also remain ventilated for a considerable amount of time: the median duration of ventilation has been reported to be five times longer in the extreme preterms born before 24 weeks of gestation compared to the ones of 25–27 weeks [7], despite the general aim towards gentle ventilation and early extubation [6].

In the very tiny infants, the ones born below 24 weeks, fine tuning of ventilation might be a crucial intervention for survival. From comparative studies from centres with favourable outcomes it appears, for example, that these centres ventilate the very immature infants either with first intention high frequency oscillation, high frequency jet or with tailored volume targeted ventilation [1]. From a pathophysiology perspective, infants born before 24 weeks of gestation have evidence of increased intrapulmonary shunting and a higher alveolar dead space compared to infants of 25–27 weeks [8]. As expected, the most immature infants also have a lower alveolar surface area. The infants in the two groups do not differ in terms of alveolar ventilation to perfusion (V_A/Q) matching which is low for both, nor do they differ in respiratory mechanics, including respiratory muscle function, compliance and resistance of the respiratory system [8]. The finding that both ventilation (alveolar dead space) and perfusion (intrapulmonary shunting) are affected, may explain why the V_A/Q is not significantly different between the two groups, but refers to an overall smaller total alveolar surface area which is available for gas exchange, as would be expected from the developmental perspective (Fig. 12.1). These pathophysiological findings might explain the relative success of some ventilatory modalities in these infants such as high frequency ventilation, as a strategy which minimises volutrauma, but might also assist in lung recruitment while avoiding perfusion abnormalities which could occur with overdistention [1].

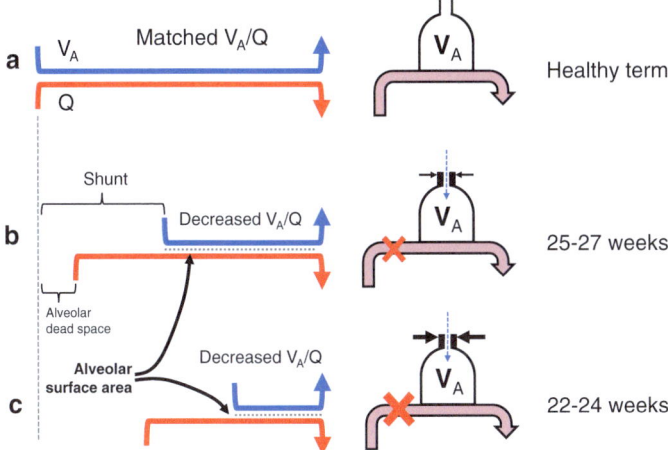

Fig. 12.1 Respiratory function at the edge of viability. Both ventilation (V_A) and perfusion (Q) are severely affected in infants of 22–24 weeks of gestation compared to extreme preterms of 25–27 weeks. The changes to ventilation and perfusion are explained by increased right-to-left shunting and increased alveolar dead space respectively. Compared to 25–27 weeks, the ratio of V_A/Q is decreased and not dissimilar in 22–24 weeks, but corresponds to a lower overall alveolar surface area

Congenital Diaphragmatic Hernia

Congenital diaphragmatic hernia (CDH) is a congenital lung disease which can serve as a model to demonstrate how a critical amount of lung parenchyma or a minimum gas-exchanging alveolar surface area are necessary for independent breathing and subsequent survival. The prognosis of CDH, including survival to discharge, is primarily determined by the extent of prenatal lung hypoplasia [9]. The degree of lung hypoplasia can be antenatally assessed with ultrasonography which holds some prognostic capability to predict survival with an average sensitivity and specificity and an area under the receiver operator characteristic curve (AUC) of 0.761 [10]. Postnatally, we have described that the adequacy of the remaining hypoplastic ipsilateral and contralateral lungs to sustain oxygenation and survival can be macroscopically assessed by the estimation of the radiographic thoracic area in plain anterio-posterior chest radiographs, which when measured on the first day of life can predict survival to discharge from neonatal care with an AUC of 0.826 [11]. These findings have also been confirmed and replicated in another two large perinatal surgical centres in different European countries which have reported similar sensitivities and specificities [12, 13]. In the population of CDH, a low V_A/Q ratio measured by the oxyhaemoglobin dissociation curve during neonatal care was associated with increased mortality, and a V_A/Q in the first 24 h of life exceeding 0.15 could predict survival to discharge with 84% sensitivity and 88% specificity and an AUC of 0.905 [14]. A markedly low V_A/Q in CDH infants is the result mostly of diffusion limitation secondary to pulmonary hypoplasia as the perfusion deficit remains unchanged during the initial days of intensive care [14] and perfusion remains persistently abnormal for the first two decades of life [15]. The anatomical dead space during the initial resuscitation in CDH infants has also been assessed using volumetric capnography, and CDH infants who did not survive had a smaller dead space than those who survived, possibly because in this case, a large dead space is a reflection of more conducting airways and less lung hypoplasia as insufficient branching of bronchioles in those with more severe lung hypoplasia would lead to fewer and smaller conducting airways and hence a lower anatomical dead space [16].

Extubation

Another clinically important critical threshold in neonatal respiratory medicine is the ability of premature infants to successfully wean off invasive mechanical ventilation and sustain independent breathing. As described earlier in this book, in utero lung development achieves an adequate interface for gas exchange following the progressive formation of alveoli and the gradually increasing inherent production of surfactant. Preterm infants also have a smaller crude volume of lungs which further impacts their capacity for adequate gas exchange [17]. The small and immature lungs of premature infants can thus negatively affect their ability to sustain independent breathing and extubate successfully. This is the rationale of why the lung volumes of premature infants have been evaluated in relation to their ability to predict successful extubation [18].

A measurement of the functional residual capacity (FRC) post-extubation, using the helium dilution technique yielded that an FRC of less than 26 ml/kg had a 71% sensitivity and 77% specificity in predicting extubation failure in premature infants, but this index did not outperform more readily available clinical parameters such as the oxygen requirement at the time point of assessment [19].

Evidently, a post-extubation measurement is not ideal in predicting the outcome of an intervention which has already happened (extubation), but unfortunately the pre-extubation measurements could not differentiate between subsequent successful or failed extubation and that gestational age had the highest AUC of all studied parameters in predicting the outcome of extubation [20]. These observations possibly underpin the two structural (functional residual capacity) and functional (oxygen requirement) components required to sustain independent breathing in preterm infants.

The two-dimensional assessment of the lung volumes obtained by measuring the lung area in chest radiographs and measuring the corresponding chest radiographic thoracic area has also been examined for the ability of this technique to predict the success of extubation. The post-extubation radiographic thoracic area in preterm infants had 100% specificity in predicting extubation failure, but similarly to the FRC studies, the gestational age and the oxygen requirement pre-extubation had a similar prognostic ability [21].

Altitude Hypoxia

As the partial pressure of oxygen decreases in higher altitudes ("high" defined as altitudes above 2500 m) [22], the respiratory system and all the peripheral tissues are exposed to a progressive degree of environmental hypoxia. Altitude hypoxia has long been an area of interest in respiratory physiology as it can serve as a fascinating model for the mechanisms of chronic hypoxia and the evolutionary adaptations of individuals and population groups exposed to chronic hypoxia [23].

The two most examined adaptations to altitude hypoxia are the ones that have developed in the Tibetan and the Andean indigenous populations. The Tibetan adaptation consists of a higher resting ventilation and a more pronounced hypoxic ventilatory response, and the Andean adaptation includes the activation of erythropoietin and the elevation in the haematocrit as compensatory mechanisms for the encountered hypoxia [24]. It is interesting that from the evolutionary perspective, the Tibetan adaptation might have appeared earlier in the evolution of man compared to the Andean one, possibly reflecting the earlier presence of humans in the area of Tibet compared to the later human migration to the Andes via the Bering strait [25]. That the response of a single individual when suddenly exposed to altitude hypoxia is a compensatory increase in the haematocrit, forms also the basis of what is known in sports medicine as *altitude training*, where competitive athletes of aerobic sports, usually runners, train for a certain period in a higher altitude compared to the one where they will engage in competitive racing, raising their haematocrit which will give them a competitive advantage compared to their co-athletes who trained at sea level [26].

Altitude hypoxia can affect the health of the newborn starting from the intrauterine period. The placental circulation demonstrates increased vascularisation and thinner villous membranes at high altitudes compared with placentas at low altitudes, possibly to facilitate the diffusion of oxygen from the maternal blood to the foetus [27, 28]. Pregnancy at high altitude also results in growth retardation and a lower birth weight [29] independently of other unfavourable demographics such as a low gestational age, the diagnosis of maternal hypertension, maternal smoking and others [30]. At birth, infants at high altitude will achieve lower arterial oxygen saturation levels compared to infants at sea level. The precise saturation values are dependent on the actual altitude with a highest arterial oxygen saturation of 87–90% in the first 48 hours at 3100 m in Colorado [31] going down to 57–75% at an extreme altitude of 4540 m in Peru [32]. Another adaptation to hypoxia is a relative increase in the proportion of foetal haemoglobin in the cord blood in high altitude compared to the percentage of foetal haemoglobin at sea level [33].

The adult ventilatory response of a lowlander (a native of a low altitude environment) during acclimatisation to high altitude is sustained hyperventilation, however in neonates this response follows a biphasic pattern with an initial hyperventilatory phase followed by a sustained decrease in the ventilation rate and a deeper breathing pattern resulting in increased oxygen extraction [34]. Morphologically, animal models have demonstrated that prenatal exposure to hypoxia is related to accelerated thinning of the alveolar walls and decreased alveolar septation, phenomena which might explain the experimental findings of a decrease in the lung elastic recoil and a larger lung volume [35]. These might also explain the larger observed lung volumes in Tibetan native highlanders [36].

The pulmonary circulation is also affected in high altitude. Chronic hypoxemia in the foetus produces an increase in the medial smooth muscle of the pulmonary arterioles, which may lead to pulmonary hypertension and increased pulmonary vasoreactivity and a slower rate of decrease in the pulmonary artery pressure following birth [31] which can lead to long term morphological impairment such as a thickened right ventricular wall [37].

The additional respiratory burden of high altitude hypoxia in preterm infants can also manifest with higher rates of bronchopulmonary dysplasia (BPD), as BPD will be more frequently diagnosed since more infants will require supplemental oxygen, and more infants will need to be discharged home on supplemental oxygen [38]. The combined effects of altitude on respiratory physiology as described above are also associated with increased morbidity in the form of more frequent and more severe respiratory events in infancy such a bronchiolitis and respiratory tract infections in infancy and early childhood [39, 40]. It is methodologically difficult however to completely separate the effects of other confounders on these outcomes, as within the same country pregnancy and birth at high altitude might be associated with unfavourable socioeconomic demographics since the more affluent families might choose to deliver at a lower altitude, in an effort to avoid these precise complications.

One pathophysiological analogue of altitude hypoxia of shorter duration which is also encountered in infants, is the exposure to the relative hypoxic environment of a commercial flight cabin where the low partial pressure of oxygen is equivalent to breathing at a fraction of inspired oxygen (FiO_2) of 0.15–0.16 at sea level [41]. As discussed earlier when we elaborated on the shape and slope of the oxyhaemoglobin dissociation curve, while this lower oxygen environment would elicit little or no clinical effect in a healthy child or adult, a steep curve which is shifted to the right (signalling V_A/Q mismatch in diseased lungs) in an ex-preterm infant might predispose the respiratory system to profound desaturations even with a modest decrease in the provided concentration of oxygen [42].

This is the physiological rationale for pre-flight testing of individuals with respiratory disease who are planning air travel [43]. The hypoxia test (fit-to-fly) is a simple method of demonstrating a prospective traveller's need for supplemental oxygen during flight and for determining in-flight oxygen requirements.

Infants with a history of neonatal lung disease studied during the second month of life, while not requiring supplemental oxygen, were exposed to 14% oxygen for a brief period of 20 minutes and were observed to have a high proportion of oxygen desaturation below 85% [44]. Interestingly, all infants had normal transcutaneous oxygen saturations ($SpO_2 > 95\%$) in room air, suggesting that baseline SpO_2 cannot discriminate the need for in-flight oxygen in this population.

Questions

Question 1: The following are true in congenital diaphragmatic hernia:
(a) Lung hypoplasia only happens at the side of the defect
(b) Hypoxaemia is uncommon
(c) Respiratory complications completely resolve after infancy
(d) Perfusion defects are common and persistent

Question 2: In high altitude environments:
(a) Adaptation to altitude hypoxia can occur via increasing the haematocrit
(b) Newborn health is not affected by intrauterine events
(c) The newborn infants will require less oxygen to maintain their saturation as they have been acclimatised
(d) The rates of bronchopulmonary dysplasia are lower compared to sea level

Question 3: The following are true regarding in-flight oxygen requirements:
(a) An extreme preterm infant that just came off supplementary oxygen and is saturating at 94% is fit to fly
(b) An ex-preterm infant on home oxygen requiring 100 cc/minute of supplemental oxygen is never fit to fly
(c) An infant saturating 90% in air with a history of congenital diaphragmatic hernia should be considered safe to fly
(d) In flight oxygen concentrations in commercial flights are equivalent to breathing 19% oxygen at sea level.
(e) A steep oxyhaemoglobin dissociation curve might predispose to severe desaturation even with mild decreases in the provided oxygen

References

1. Sindelar R, Nakanishi H, Stanford AH, Colaizy TT, Klein JM. Respiratory management for extremely premature infants born at 22 to 23 weeks of gestation in proactive centers in Sweden, Japan, and USA. Semin Perinatol. 2022;46(1):151540. Epub 2021/12/08. https://doi.org/10.1016/j.semperi.2021.151540.
2. Ahmad KA, Frey CS, Fierro MA, Kenton AB, Placencia FX. Two-year neurodevelopmental outcome of an infant born at 21 weeks' 4 days' gestation. Pediatrics. 2017;140(6). Epub 2017/11/04. https://doi.org/10.1542/peds.2017-0103.
3. Sweet DG, Carnielli VP, Greisen G, Hallman M, Klebermass-Schrehof K, Ozek E, et al. European consensus guidelines on the management of respiratory distress syndrome: 2022 update. Neonatology. 2023;120(1):3–23. Epub 2023/03/03. https://doi.org/10.1159/000528914.

4. Jobe AH, Ikegami M. Mechanisms initiating lung injury in the preterm. Early Hum Dev. 1998;53(1):81–94. Epub 1999/04/08. https://doi.org/10.1016/s0378-3782(98)00045-0.
5. Jobe AH. Mechanisms of lung injury and bronchopulmonary dysplasia. Am J Perinatol. 2016;33(11):1076–8. https://doi.org/10.1055/s-0036-1586107.
6. Norman M, Jonsson B, Soderling J, Bjorklund LJ, Hakansson S. Patterns of respiratory support by gestational age in very preterm infants. Neonatology. 2023;120(1):142–52. Epub 2022/12/13. https://doi.org/10.1159/000527641.
7. Dassios T, Williams EE, Hickey A, Bunce C, Greenough A. Bronchopulmonary dysplasia and postnatal growth following extremely preterm birth. Arch Dis Child Fetal Neonatal Ed. 2021;106(4):386–91. Epub 2020/12/19. https://doi.org/10.1136/archdischild-2020-320816.
8. Dassios T, Sindelar R, Williams E, Kaltsogianni O, Greenough A. Invasive ventilation at the boundary of viability: a respiratory pathophysiology study of infants born between 22 and 24 weeks of gestation. Respir Physiol Neurobiol. 2025;331:104339. Epub 2024/09/06. https://doi.org/10.1016/j.resp.2024.104339.
9. Ackerman KG, Pober BR. Congenital diaphragmatic hernia and pulmonary hypoplasia: new insights from developmental biology and genetics. Am J Med Genet C Semin Med Genet. 2007;145C(2):105–8. Epub 2007/04/17. https://doi.org/10.1002/ajmg.c.30133.
10. Jani J, Nicolaides KH, Keller RL, Benachi A, Peralta CF, Favre R, et al. Observed to expected lung area to head circumference ratio in the prediction of survival in fetuses with isolated diaphragmatic hernia. Ultrasound Obstet Gynecol. 2007;30(1):67–71. Epub 2007/06/26. https://doi.org/10.1002/uog.4052.
11. Dassios T, Ali K, Makin E, Bhat R, Krokidis M, Greenough A. Prediction of mortality in newborn infants with severe congenital diaphragmatic hernia using the chest radiographic thoracic area. Pediatr Crit Care Med. 2019;20(6):534–9. Epub 2019/03/30. https://doi.org/10.1097/PCC.0000000000001912.
12. Weis M, Burhany S, Perez Ortiz A, Nowak O, Hetjens S, Zahn K, et al. The chest radiographic thoracic area can serve as a prediction marker for morbidity and mortality in infants with congenital diaphragmatic hernia. Front Pediatr. 2021;9:740941. Epub 2022/01/11. https://doi.org/10.3389/fped.2021.740941.
13. Amodeo I, Pesenti N, Raffaeli G, Macchini F, Condo V, Borzani I, et al. NeoAPACHE II. Relationship between radiographic pulmonary area and pulmonary hypertension, mortality, and hernia recurrence in newborns with CDH. Front Pediatr. 2021;9:692210. Epub 2021/07/30. https://doi.org/10.3389/fped.2021.692210.

14. Dassios T, Shareef Arattu Thodika FM, Williams E, Davenport M, Nicolaides KH, Greenough A. Ventilation-to-perfusion relationships and right-to-left shunt during neonatal intensive care in infants with congenital diaphragmatic hernia. Pediatr Res. 2022;92:1657–62. Epub 2022/03/21. https://doi.org/10.1038/s41390-022-02001-2.
15. Dao DT, Kamran A, Wilson JM, Sheils CA, Kharasch VS, Mullen MP, et al. Longitudinal analysis of ventilation perfusion mismatch in congenital diaphragmatic hernia survivors. J Pediatr. 2020;219:160–6 e2. Epub 2019/11/11. https://doi.org/10.1016/j.jpeds.2019.09.053.
16. Williams EE, Dassios T, Murthy V, Greenough A. Anatomical deadspace during resuscitation of infants with congenital diaphragmatic hernia. Early Hum Dev. 2020;149:105150. Epub 2020/08/11. https://doi.org/10.1016/j.earlhumdev.2020.105150.
17. De Paepe ME, Shapiro S, Hansen K, Gundogan F. Postmortem lung volume/body weight standards for term and preterm infants. Pediatr Pulmonol. 2014;49(1):60–6. Epub 2013/09/17. https://doi.org/10.1002/ppul.22818.
18. Shalish W, Latremouille S, Papenburg J, Sant'Anna GM. Predictors of extubation readiness in preterm infants: a systematic review and meta-analysis. Arch Dis Child Fetal Neonatal Ed. 2019;104(1):F89–97. https://doi.org/10.1136/archdischild-2017-313878.
19. Dimitriou G, Greenough A, Laubscher B. Lung volume measurements immediately after extubation by prediction of "extubation failure" in premature infants. Pediatr Pulmonol. 1996;21(4):250–4. Epub 1996/04/01. https://doi.org/10.1002/(SICI)1099-0496(199604)21:4<250::AID-PPUL9>3.0.CO;2-S.
20. Kavvadia V, Greenough A, Dimitriou G. Prediction of extubation failure in preterm neonates. Eur J Pediatr. 2000;159(4):227–31. Epub 2000/05/02. https://doi.org/10.1007/s004310050059.
21. Dimitriou G, Greenough A. Computer assisted analysis of the chest radiograph lung area and prediction of failure of extubation from mechanical ventilation in preterm neonates. Br J Radiol. 2000;73(866):156–9. Epub 2000/07/08. https://doi.org/10.1259/bjr.73.866.10884728.
22. Boning D. Altitude and hypoxia training – a short review. Int J Sports Med. 1997;18(8):565–70. Epub 1998/01/27. https://doi.org/10.1055/s-2007-972682.
23. West JB. High-altitude medicine. Am J Respir Crit Care Med. 2012;186(12):1229–37. Epub 2012/10/30. https://doi.org/10.1164/rccm.201207-1323CI.
24. Beall CM. Tibetan and Andean contrasts in adaptation to high-altitude hypoxia. Adv Exp Med Biol. 2000;475:63–74. Epub 2000/06/13. https://doi.org/10.1007/0-306-46825-5_7.
25. Beall CM. Two routes to functional adaptation: Tibetan and Andean high-altitude natives. Proc Natl Acad Sci U S Am. 2007;104 Suppl 1(Suppl 1):8655–60. Epub 2007/05/15. https://doi.org/10.1073/pnas.0701985104.

26. Levine BD, Stray-Gundersen J. A practical approach to altitude training: where to live and train for optimal performance enhancement. Int J Sports Med. 1992;13(Suppl 1):S209–12. Epub 1992/10/01. https://doi.org/10.1055/s-2007-1024642.
27. Tissot van Patot MC, Bendrick-Peart J, Beckey VE, Serkova N, Zwerdlinger L. Greater vascularity, lowered HIF-1/DNA binding, and elevated GSH as markers of adaptation to in vivo chronic hypoxia. Am J Physiol Lung Cell Mol Physiol. 2004;287(3):L525–32. Epub 2004/05/11. https://doi.org/10.1152/ajplung.00203.2003.
28. Ahrens S, Singer D. Placental adaptation to hypoxia: the case of high-altitude pregnancies. Int J Environ Res Public Health. 2025;22(2). Epub 2025/02/26. https://doi.org/10.3390/ijerph22020214.
29. Unger C, Weiser JK, McCullough RE, Keefer S, Moore LG. Altitude, low birth weight, and infant mortality in Colorado. JAMA. 1988;259(23):3427–32. Epub 1988/06/17
30. Jensen GM, Moore LG. The effect of high altitude and other risk factors on birthweight: independent or interactive effects? Am J Public Health. 1997;87(6):1003–7. Epub 1997/06/01. https://doi.org/10.2105/ajph.87.6.1003.
31. Niermeyer S, Shaffer EM, Thilo E, Corbin C, Moore LG. Arterial oxygenation and pulmonary arterial pressure in healthy neonates and infants at high altitude. J Pediatr. 1993;123(5):767–72. Epub 1993/11/01. https://doi.org/10.1016/s0022-3476(05)80857-1.
32. Gamboa R, Marticorena E. [Pulmonary arterial pressure in newborn infants in high altitude]. Arch Inst Biol Andina. 1971;4(2):55–66. Epub 1971/05/01. Presion arterial pulmonar en el recien nacido en las grandes alturas.
33. Ballew C, Haas JD. Hematologic evidence of fetal hypoxia among newborn infants at high altitude in Bolivia. Am J Obstet Gynecol. 1986;155(1):166–9. Epub 1986/07/01.10.1016/0002-9378(86)90104-3
34. Niermeyer S. Cardiopulmonary transition in the high altitude infant. High Alt Med Biol. 2003;4(2):225–39. Epub 2003/07/12. https://doi.org/10.1089/152702903322022820.
35. Massaro GD, Olivier J, Dzikowski C, Massaro D. Postnatal development of lung alveoli: suppression by 13% O2 and a critical period. Am J Physiol. 1990;258(6 Pt 1):L321–7. Epub 1990/06/01. https://doi.org/10.1152/ajplung.1990.258.6.L321.
36. Wu T, Kayser B. High altitude adaptation in Tibetans. High Alt Med Biol. 2006;7(3):193–208. Epub 2006/09/19. https://doi.org/10.1089/ham.2006.7.193.
37. Aparicio Otero O, Romero Gutierrez F, Harris P, Anand I. Echocardiography shows persistent thickness of the wall of the right ventricle in infants at high altitude. Cardioscience. 1991;2(1):63–9. Epub 1991/03/01

38. Fernandez CL, Fajardo CA, Favareto MV, Hoyos A, Jijon-Letort FX, Carrera MS, et al. Oxygen dependency as equivalent to bronchopulmonary dysplasia at different altitudes in newborns ⩽ 1500 g at birth from the SIBEN network. J Perinatol. 2014;34(7):538–42. Epub 2014/04/05. https://doi.org/10.1038/jp.2014.46.
39. Lowther SA, Shay DK, Holman RC, Clarke MJ, Kaufman SF, Anderson LJ. Bronchiolitis-associated hospitalizations among American Indian and Alaska Native children. Pediatr Infect Dis J. 2000;19(1):11–7. Epub 2000/01/22. https://doi.org/10.1097/00006454-200001000-00004.
40. Duke T, Mgone J, Frank D. Hypoxaemia in children with severe pneumonia in Papua New Guinea. Int J Tuberc Lung Dis. 2001;5(6):511–9. Epub 2001/06/21
41. Samuels MP. The effects of flight and altitude. Arch Dis Childhood. 2004;89(5):448–55. Epub 2004/04/23. https://doi.org/10.1136/adc.2003.031708.
42. Jones JG, Lockwood GG, Fung N, Lasenby J, Ross-Russell RI, Quine D, et al. Influence of pulmonary factors on pulse oximeter saturation in preterm infants. Arch Dis Child Fetal Neonatal Ed. 2016;101(4):F319–22. https://doi.org/10.1136/archdischild-2015-308675.
43. Ahmedzai S, Balfour-Lynn IM, Bewick T, Buchdahl R, Coker RK, Cummin AR, et al. Managing passengers with stable respiratory disease planning air travel: British Thoracic Society recommendations. Thorax. 2011;66(Suppl 1):i1–30. Epub 2011/10/04. https://doi.org/10.1136/thoraxjnl-2011-200295.
44. Udomittipong K, Stick SM, Verheggen M, Oostryck J, Sly PD, Hall GL. Pre-flight testing of preterm infants with neonatal lung disease: a retrospective review. Thorax. 2006;61(4):343–7. Epub 2006/02/02. https://doi.org/10.1136/thx.2005.048769.

Children and Adults with a History of Neonatal Lung Disease

13

While measurement of lung function is not always feasible in the newborn as most tests are volitional and require the interactive cooperation of the subject undertaking the test, traditional measurements of lung function and spirometry are possible in older children and adults.

There are two main epidemiological observations which are useful when considering lung function in older children and adults with a history of neonatal lung disease.

Known Unknowns

Neonatal respiratory care has changed dramatically over the past decades and continues to change, originally with antenatal corticosteroids and postnatal surfactant in the 1990s, followed by similar later breakthroughs such as high frequency oscillation and volume targeted ventilation, avoidance of invasive ventilation and more sophisticated methods of non-invasive support which coupled with overall improved extra-respiratory care have altered the epidemiology of BPD and the population of infants who survive from extreme prematurity usually with severe lung disease [1]. This new population, consists of even more immature infants at birth, who have avoided the bland trauma of excessive oxygen toxicity and barotrauma, but because they were born earlier in gestation, they have more severe lung disease as their lung growth was disrupted at an earlier developmental stage [2]. The full scale of these developments has not been fully revealed yet, as these newborns are only now reaching adolescence and young adulthood while in the same period the neonatal care and thresholds of survival keep improving.

The practical consequence of this epidemiological phenomenon is that we have simply not seen the full picture of adult BPD yet, which is still actively evolving. We do not know how these infants will be as adults nor do we have any real data on how they will age. We are only starting to measure them at an adult age, thanks mainly to some medical teams around the world who have painstakingly maintained contact over decades with the same invaluable cohorts, have succeeded in getting them followed up and have their lung function measured serially over the span of more than three decades [3, 4].

Unknown Unknowns

In classical Greek theatre, the literary technique when the audience was made aware of a future development in the plot while the actor was not yet aware of it, was called *dramatic irony*. There is potentially a similar situation currently unfolding in our time with the progression of lung disease in adults who as newborns had significant respiratory disease and especially the preterm ones. These individuals have been diagnosed with lung disease as newborns, they then gradually grew out of their lung disease as they became asymptomatic and did not require regular medications and support. Consequently they stopped being followed up and were discharged as healthy individuals, but are potentially due to return to adult pulmonology care at an earlier age than expected due to the progression of their lung disease over time, which remains currently subclinical.

It is useful to remember, that one does not need all the available lung reserves for performing their everyday activities and it is possible that lung function starts off by being very abnormal in the very early days of neonatal lung disease, but then as these infants grow and overcome some degree of the severity of their lung disease, there follows a period free of symptoms or limitations. This means that they operate above the "symptoms threshold", which is well below the 100% predicted values of spirometric lung function at a population level. For example, if a young adult who had BPD as a newborn, at 25 years of age has a FEV_1 of 85%, this would not affect them in their everyday activities and the quality of life would not be severely disturbed. Over time, however, lung function would follow a predefined trajectory where FEV_1 increases in childhood and young adulthood until the mid-twenties and then starts naturally to follow a slow decline, in such a slow way that healthy adults who don't smoke or develop other diseases as children or adults will not hit the "symptoms threshold" during their expected lifetime (Fig. 13.1). Individuals, however, who as newborns had severe lung disease or BPD, will follow a different course. They will start by abnormal values

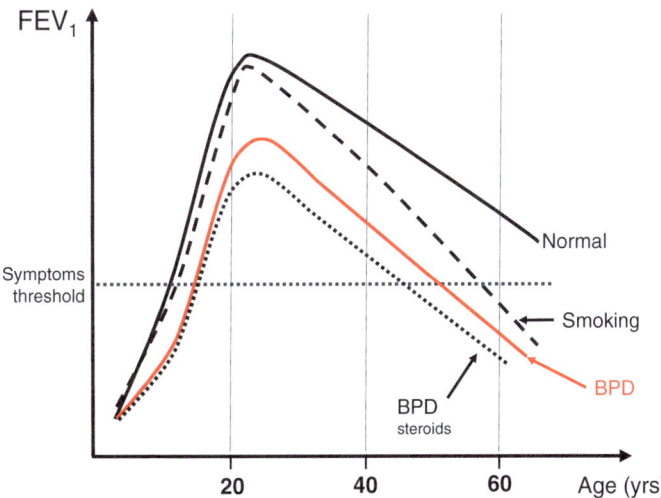

Fig. 13.1 Lung function trajectories across the lifespan. The curve in healthy individuals follows a more gradual decline than the curve in smoking individuals. The BPD curve does not reach the maximum value compared to the normal curve. With a potential faster rate of decline, the BPD curve might reach the symptoms threshold at an earlier age

which will track at a lower trajectory, might not achieve their full lung function potential and will eventually decline in such a way, that their lung function will hit the "symptoms threshold" at a much earlier age (the estimated guess is mid-forties) and will present with symptomatic lung disease despite not having a history of asthma, chronic obstructive pulmonary disease (COPD) or any other condition that could explain this clinical picture other than the history of extreme prematurity and neonatal lung disease [5].

In practice, this means that these individuals will be discharged from follow up in childhood as they will then be symptom free and will reappear some decades later as adult patients with symptoms that resemble COPD. The rate of lung function decline can be accelerated by some parameters such as typically smoking, so it is very important that these individuals are counselled to avoid this habit [6]. In the neonatal population in specific, it appears that administration of postnatal steroids is also associated with a lower maximum lung function potential and a faster rate of decline, as corticosteroids further aggravate lung growth retardation or arrest which is the primary pathophysiological hallmark of BPD in the first place [7]. It is also possible that structured aerobic exercise might benefit these individuals as an intended planned intervention to delay the rate of lung function decline or improve lung function, as demonstrated by other chronic paediatric lung diseases such as Cystic Fibrosis [8].

The Respiratory BPD Phenotype in Older Children and Adults

Children and adults with a history of BPD were initially thought to develop primarily an obstructive pattern of respiratory disease, but maybe this is because these were the only available tests we could use during their follow up at the time. These tests were mainly related to airway function, such as the forced expiratory volumes and flows. Mirroring the changing epidemiology of BPD, it has gradually become clearer that in later life the survivors of BPD exhibit a composite disease phenotype with both

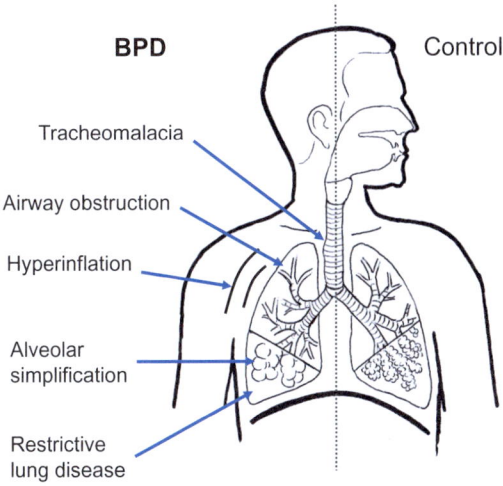

Fig. 13.2 The adult BPD respiratory phenotype. Adult survivors of BPD exhibit evidence of tracheomalacia, airway obstruction, hyperinflation, alveolar simplification with reduced exercise capacity and restricted lung volumes

obstructive and restrictive elements (Fig. 13.2). These individuals are often symptomatic with symptoms suggestive of airway obstruction, such as wheezing and shortness of breath, although the underlying pathology in BPD survivors is often fixed airway obstruction due to structural damage rather than reversible inflammation, as seen in asthma. They also manifest with symptoms suggestive of restrictive lung disease such as exercise limitation which is possibly the combined result of a reduced alveolar surface area as well as airway limitation during exercise [9].

Obstructive Lung Disease

With regards to evidence describing an obstructive type of lung disease, adult survivors of BPD have been reported to have decreased forced expiratory volume in 1 second (FEV_1) and the ratio of FEV_1 to the forced vital capacity (FEV_1/FVC ratio) on spirometry [9]. This obstruction is usually less responsive to bronchodilator therapy as it is more fixed and structural in origin due to impaired development of the small airways [10]. Other than this fixed developmental component, the airways of BPD survivors have a degree of inflammation and immune dysregulation which further aggravates airway obstruction [11]. Whether the adult survivors of BPD can be formally labelled as a subcategory of COPD as they meet many of the diagnostic criteria, is a matter of active scientific contention, which might be mostly academic in nature and will possibly have a limited effect on their management approach [10].

Hyperinflation

A certain degree of hyperinflation secondary to air trapping is also present in longitudinal studies of lung function in BPD survivors [4].

Restrictive Lung Disease

Children and adults with BPD have also evidence of restrictive lung disease with low FVC, diffusing capacity of the lungs for carbon monoxide (DLCO) and decreased exercise capacity [12].

Exercise Capacity

Exercise testing can reveal abnormalities of the cardiorespiratory system which are not apparent in everyday life or by conventional tests of lung function such as spirometry.

Adult survivors of prematurity have reduced exercise capacity and exercise tolerance [13]. The lung function findings include a lower maximum rate of oxygen consumption attainable during physical exertion (VO_2 max), decreased diffusion capacity, and impaired lung flow due to fixed obstruction [14].

Pulmonary Hypertension

Some adult survivors of BPD exhibit increased pulmonary vascular resistance which can increase the right ventricular afterload and cause right ventricular dysfunction [15]. At a population level adult survivors of prematurity have a three to five-fold increase in the risk for pulmonary hypertension [16, 17].

Dysynaptic Lung Growth

Pulmonary dysynapsis describes the condition where there is imbalanced growth between the airways and the pulmonary parenchyma. In the case of the adult BPD phenotype this imbalance refers to narrower airways relative to more preserved lung size which can lead to less efficient air flow through the airways. Dysanapsis can clinically manifest with airflow limitation, air trapping and impaired exercise capacity.

Given the above it is clinically important to investigate potential reversible cause of obstruction and administer treatment when indicated potentially with bronchodilators, respiratory muscle training and pulmonary physiotherapy and rehabilitation programs which can probably improve air trapping and expiratory flow limitation by augmenting expiratory muscle strength [9].

Questions

Question 1: Regarding lung health in adults with a history of neonatal lung disease:
 (a) A history of prematurity is associated with a higher maximum lung function potential
 (b) Newborns exposed to systemic postnatal steroids have preserved lung function in adulthood compared to the infants who did not receive steroids
 (c) Smoking accelerates lung function decline
 (d) Adults with a history of BPD should avoid exercise as they might not be fit enough

Question 2: The following are characteristics of adults with a history of BPD
(a) Higher FEV/FVC ratio
(b) Low FEV_1
(c) Lower functional residual capacity and residual volume
(d) Increased exercise capacity at baseline
(e) Right ventricular atrophy

References

1. Day CL, Ryan RM. Bronchopulmonary dysplasia: new becomes old again! Pediatr Res. 2017;81(1–2):210–3. Epub 2016/10/27. https://doi.org/10.1038/pr.2016.201.
2. Jobe AH. The new bronchopulmonary dysplasia. Curr Opin Pediatr. 2011;23(2):167–72. Epub 2010/12/21. https://doi.org/10.1097/MOP.0b013e3283423e6b.
3. Moschino L, Stocchero M, Filippone M, Carraro S, Baraldi E. Longitudinal assessment of lung function in survivors of bronchopulmonary dysplasia from birth to adulthood. The Padova BPD study. Am J Respir Crit Care Med. 2018;198(1):134–7. Epub 2018/02/23. https://doi.org/10.1164/rccm.201712-2599LE.
4. Harris C, Morris S, Lunt A, Peacock J, Greenough A. Influence of bronchopulmonary dysplasia on lung function in adolescents who were born extremely prematurely. Pediatr Pulmonol. 2022;57(12):3151–7. Epub 2022/09/14. https://doi.org/10.1002/ppul.26151.
5. Bolton CE, Bush A, Hurst JR, Kotecha S, McGarvey L. Lung consequences in adults born prematurely. Thorax. 2015;70(6):574–80. Epub 2015/04/01. https://doi.org/10.1136/thoraxjnl-2014-206590.
6. Burchfiel CM, Marcus EB, Curb JD, Maclean CJ, Vollmer WM, Johnson LR, et al. Effects of smoking and smoking cessation on longitudinal decline in pulmonary function. Am J Respir Crit Care Med. 1995;151(6):1778–85. Epub 1995/06/01. https://doi.org/10.1164/ajrccm.151.6.7767520.
7. Harris C, Bisquera A, Zivanovic S, Lunt A, Calvert S, Marlow N, et al. Postnatal dexamethasone exposure and lung function in adolescents born very prematurely. PLoS One. 2020;15(8):e0237080. https://doi.org/10.1371/journal.pone.0237080.
8. Kriemler S, Kieser S, Junge S, Ballmann M, Hebestreit A, Schindler C, et al. Effect of supervised training on FEV1 in cystic fibrosis: a randomised controlled trial. J Cyst Fibros. 2013;12(6):714–20. Epub 2013/04/17. https://doi.org/10.1016/j.jcf.2013.03.003.

9. Wozniak PS, Makhoul L, Botros MM. Bronchopulmonary dysplasia in adults: exploring pathogenesis and phenotype. Pediatr Pulmonol. 2024;59(3):540–51. Epub 2023/12/05. https://doi.org/10.1002/ppul.26795.
10. Bardsen T, Roksund OD, Eagan TML, Hufthammer KO, Benestad MR, Clemm HSH, et al. Impaired lung function in extremely preterm-born adults in their fourth decade of life. Am J Respir Crit Care Med. 2023;208(4):493–5. Epub 2023/05/16. https://doi.org/10.1164/rccm.202303-0448LE.
11. Um-Bergstrom P, Pourbazargan M, Brundin B, Strom M, Ezerskyte M, Gao J, et al. Increased cytotoxic T-cells in the airways of adults with former bronchopulmonary dysplasia. Eur Respir J. 2022;60(3). Epub 2022/02/26. https://doi.org/10.1183/13993003.02531-2021.
12. Cassady SJ, Lasso-Pirot A, Deepak J. Phenotypes of Bronchopulmonary Dysplasia in Adults. Chest. 2020;158(5):2074–81. Epub 2020/06/01. https://doi.org/10.1016/j.chest.2020.05.553.
13. Lovering AT, Elliott JE, Laurie SS, Beasley KM, Gust CE, Mangum TS, et al. Ventilatory and sensory responses in adult survivors of preterm birth and bronchopulmonary dysplasia with reduced exercise capacity. Ann Am Thorac Soc. 2014;11(10):1528–37. https://doi.org/10.1513/AnnalsATS.201312-466OC.
14. Yang J, Epton MJ, Harris SL, Horwood J, Kingsford RA, Troughton R, et al. Reduced exercise capacity in adults born at very low birth weight: a population-based cohort study. Am J Respir Crit Care Med. 2022;205(1):88–98. Epub 2021/09/10. https://doi.org/10.1164/rccm.202103-0755OC.
15. Goss KN, Beshish AG, Barton GP, Haraldsdottir K, Levin TS, Tetri LH, et al. Early pulmonary vascular disease in young adults born preterm. Am J Respir Crit Care Med. 2018;198(12):1549–58. Epub 2018/06/27. https://doi.org/10.1164/rccm.201710-2016OC.
16. Naumburg E, Soderstrom L, Huber D, Axelsson I. Risk factors for pulmonary arterial hypertension in children and young adults. Pediatr Pulmonol. 2017;52(5):636–41. Epub 2016/11/02. https://doi.org/10.1002/ppul.23633.
17. Naumburg E, Axelsson I, Huber D, Soderstrom L. Some neonatal risk factors for adult pulmonary arterial hypertension remain unknown. Acta Paediatr. 2015;104(11):1104–8. Epub 2015/09/09. https://doi.org/10.1111/apa.13205.

Lung Function Tests in Neonates

14

Lung function testing in cooperative subjects includes forced expiratory manoeuvres which are the most common spirometric tests and are offered widely at an inpatient and outpatient basis. Unfortunately, since infants cannot follow commands and do volitional tests, pulmonary function testing in neonates will alternatively utilise more complex methods which are rarely used in clinical practice and are only offered at specialised centres. Lung function testing in neonatology is a broad subject [1] which cannot be exhausted in this short chapter, where the aim is to briefly summarise physiological findings in health and disease and novel promising methodologies which might find direct clinical applicability for the wider neonatal community in the near future.

One example of a relatively cumbersome methodology includes the acquisition of forced expiratory flow volume loops, which can be obtained by externally applied pressure to the chest and abdomen to mimic a forced expiration. Studies on infants and young children with a history of neonatal acute or chronic lung disease have demonstrated that forced flows remain persistently decreased through the first three years of life [2].

Abnormal airway function can also be assessed during tidal breathing by the time to peak tidal expiratory flow as a proportion of the total expiratory time (T_{PTEF}/T_E). The advantage of tidal breathing is that tidal parameters correspond to quiet normal breathing and these parameters can be recorded without exerting

© The Author(s), under exclusive license to Springer Nature Switzerland AG 2025
T. Dassios, *Clinical Respiratory Physiology of the Newborn*, In Clinical Practice, https://doi.org/10.1007/978-3-032-05738-9_14

some external force to mimic a maximal expiration, while they can also be non-invasively derived via respiratory inductive plethysmography or flow sensors. In concept, low values of T_{PTEF}/T_E (Fig. 14.1) describe abnormal airway function and when measured early in infancy have been linked with the development of later wheezing [3]. When measured, however, earlier in the first week of life in term infants, this index has limited ability for the prediction of future respiratory disease [4]. Tidal breathing measurements are also different in infants with BPD compared to term controls with a shorter inspiratory rate and higher respiratory rate in BPD, but the expiratory flow parameters could not discriminate between diseased and control infants [5]. When the T_{PTEF}/T_E, however, is measured earlier in life in ventilated preterm infants, it has been found to be significantly lower compared to term infants, as well as in prematurely born infants who later develop BPD compared to those who do not [6].

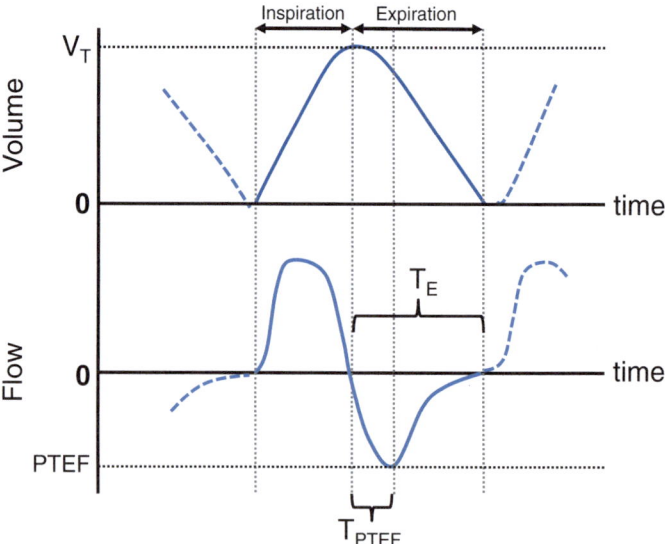

Fig. 14.1 Tidal breathing graphs of volume and flow over time. V_T tidal volume, PTEF peak tidal expiratory flow, T_E expiratory time, T_{PTEF} time to peak tidal expiratory flow

The functional residual capacity (FRC) can also be measured in neonates using plethysmography and gas dilution/washout techniques. The FRC describes the resting lung volume at end-expiration, and is the only static lung volume that can be readily assessed in non-cooperative infants and very young children [2]. Acute neonatal lung disease is characterised by severely reduced FRC values, with treatments such as continuous positive airway pressure and endotracheal surfactant aimed at optimising lung recruitment. While FRC may remain reduced in established chronic lung disease of infancy, more commonly it gradually increases due to hyperinflation, with or without gas-trapping, secondary to airway obstruction [7].

A novel and more accessible method to estimate the FRC is via *electrical impedance tomography* (EIT) which uses the term end-expiratory lung volume (EELV) as an equivalent to FRC. Electrical impedance tomography is a non-invasive technique of generating real-time information on lung morphology and aeration using measurements of the spatial distributions of resistivity (impedance to the flow of electric current) of tissues obtained from electrodes attached around the circumference of the chest [8]. Lung volume increased in response to surfactant treatment in preterm infants with RDS on mechanical ventilation, and there was increased ventilation in the dorsal lung and more homogenous right-to-left ventilation distribution post surfactant treatment [9]. EIT has also been used to assess differences between modes of invasive and non-invasive ventilation. Tidal volumes have been shown to be comparable on nasal high frequency oscillation compared to nasal continuous positive airway pressure (CPAP) in preterm infants [10]. Preterm neonates on invasive ventilation versus CPAP or high flow nasal cannula therapy, exhibited higher end-expiratory lung impedance with invasive ventilation but greater ventilation in the dependent lung areas with the non-invasive methods [11]. Similarly, EIT measurements of the change in end-expiratory lung impedance have demonstrated that prone positioning enhances dorsal lung aeration, which can reduce atelectasis, while supine positioning promotes more symmetrical ventilation [8].

Titration of positive end-expiratory pressure (PEEP) has also been studied using EIT with an aim to minimize ventilation inhomogeneity during spontaneous assisted ventilation and as a guide to minimize overdistention and atelectasis in infants with BPD [12].

Lung ultrasound has also gained popularity in the recent years, especially since neonatal health professionals are already broadly familiar with the technology in the context mainly of head and heart scanning. Using such bedside ultrasonographic assessments, lung function can be dynamically and serially monitored in a variety of conditions such as in the transient tachypnoea of the newborn, respiratory distress syndrome, meconium aspiration syndrome, acute respiratory distress syndrome, pneumonia and air leak syndromes [13].

Another emerging methodology to non-invasively assess neonatal lung function at the cot side is respiratory oscillometry, which is a technique that measures respiratory mechanics passively in spontaneous breathing subjects. Oscillatory pressure waves are applied to the respiratory system and the resulting pressure and flow changes are measured to derive respiratory impedance (Zrs), which includes resistance (Rrs) and reactance (Xrs) [14]. This technique describes the mechanical properties of the lungs, such as the airway resistance and compliance. Although there is significant intra-breath fluctuations in Rrs and Xrs within the breathing cycle [15] studies using the forced oscillation technique (FOT), have found that infants with transient tachypnoea of the newborn had a higher resistance and lower reactance on the first day of life compared to healthy controls [16]. The Xrs assessed within two hours of life in preterm infants can predict the need for surfactant and the duration of respiratory support [17].

Given that the main chronic pathophysiological characteristic of respiratory morbidity following premature birth is the arrest or severe retardation of alveolar growth, we have also recently studied and validated a non-invasive index of the alveolar surface area which can be derived by paired measurements of transcutaneous oxygen saturations and the corresponding provided fraction of inspired oxygen [18]. Although, clearly, more studies will be needed to replicate these findings before the index is adopted for clinical use, this index has been shown to predict the need for home oxygen and was significantly abnormal in infants with neurodevelopmental impairment at two years of corrected age [19].

Question

Question 1: Regarding lung function testing in neonates
 (a) Infants cannot execute commands, however neonatal lung function testing is widely available in all neonatal units
 (b) Flow volume loops rapidly become normal following discharge from neonatal care
 (c) In electrical impedance tomography there is ionising radiation
 (d) Electrical impedance tomography can provide real time information on regional ventilation
 (e) Forced oscillation requires muscle relaxation in infants

References

1. Go MA, MacDonald KD, Durand M, McEvoy CT. Pulmonary function tests in the neonatal intensive care unit and beyond: a clinical review. J Perinatol. 2025. Epub 2025/03/01. https://doi.org/10.1038/s41372-025-02243-y.
2. Lum S, Hulskamp G, Merkus P, Baraldi E, Hofhuis W, Stocks J. Lung function tests in neonates and infants with chronic lung disease: forced expiratory maneuvers. Pediatr Pulmonol. 2006;41(3):199–214. Epub 2005/11/17. https://doi.org/10.1002/ppul.20320.
3. Martinez FD, Morgan WJ, Wright AL, Holberg CJ, Taussig LM. Diminished lung function as a predisposing factor for wheezing respiratory illness in infants. N Engl J Med. 1988;319(17):1112–7. Epub 1988/10/27. https://doi.org/10.1056/NEJM198810273191702.
4. Yuksel B, Greenough A, Giffin F, Nicolaides KH. Tidal breathing parameters in the first week of life and subsequent cough and wheeze. Thorax. 1996;51(8):815–8. Epub 1996/08/01. https://doi.org/10.1136/thx.51.8.815.
5. Schmalisch G, Wilitzki S, Wauer RR. Differences in tidal breathing between infants with chronic lung diseases and healthy controls. BMC Pediatr. 2005;5:36. Epub 2005/09/10. https://doi.org/10.1186/1471-2431-5-36.

6. Dassios T, Kaltsogianni O, Greenough A. Determinants of pulmonary dead space in ventilated newborn infants. Early Hum Dev. 2017;108:29–32. https://doi.org/10.1016/j.earlhumdev.2017.03.011.
7. Maynard V, Bignall S, Kitchen S. Effect of positioning on respiratory synchrony in non-ventilated pre-term infants. Physiother Res Int. 2000;5(2):96–110.
8. Ako AA, Ismaiel A, Rastogi S. Electrical impedance tomography in neonates: a review. Pediatr Res. 2025. Epub 2025/02/23. https://doi.org/10.1038/s41390-025-03929-x.
9. Chatziioannidis I, Samaras T, Mitsiakos G, Karagianni P, Nikolaidis N. Assessment of lung ventilation in infants with respiratory distress syndrome using electrical impedance tomography. Hippokratia. 2013;17(2):115–9. Epub 2014/01/01
10. Gaertner VD, Waldmann AD, Davis PG, Bassler D, Springer L, Thomson J, et al. Transmission of oscillatory volumes into the preterm lung during noninvasive high-frequency ventilation. Am J Respir Crit Care Med. 2021;203(8):998–1005. Epub 2020/10/24. https://doi.org/10.1164/rccm.202007-2701OC.
11. Virsilas E, Valiulis A, Kubilius R, Peciuliene S, Liubsys A. Respiratory support effects over time on regional lung ventilation assessed by electrical impedance tomography in premature infants. Medicina (Kaunas). 2024;60(3). Epub 2024/03/28. https://doi.org/10.3390/medicina60030494.
12. Zhao Z, Steinmann D, Frerichs I, Guttmann J, Moller K. PEEP titration guided by ventilation homogeneity: a feasibility study using electrical impedance tomography. Crit Care. 2010;14(1):R8. Epub 2010/02/02. https://doi.org/10.1186/cc8860.
13. Raimondi F, Yousef N, Migliaro F, Capasso L, De Luca D. Point-of-care lung ultrasound in neonatology: classification into descriptive and functional applications. Pediatr Res. 2021;90(3):524–31. Epub 2018/08/22. https://doi.org/10.1038/s41390-018-0114-9.
14. Kaminsky DA, Simpson SJ, Berger KI, Calverley P, de Melo PL, Dandurand R, et al. Clinical significance and applications of oscillometry. Eur Respir Rev. 2022;31(163). Epub 2022/02/11. https://doi.org/10.1183/16000617.0208-2021.
15. Radics BL, Gyurkovits Z, Makan G, Gingl Z, Czovek D, Hantos Z. Respiratory oscillometry in newborn infants: conventional and intrabreath approaches. Front Pediatr. 2022;10:867883. Epub 2022/04/22. https://doi.org/10.3389/fped.2022.867883.
16. Klinger AP, Travers CP, Martin A, Kuo HC, Alishlash AS, Harris WT, et al. Non-invasive forced oscillometry to quantify respiratory mechanics in term neonates. Pediatr Res. 2020;88(2):293–9. Epub 2020/01/15. https://doi.org/10.1038/s41390-020-0751-7.

17. Lavizzari A, Veneroni C, Beretta F, Ottaviani V, Fumagalli C, Tossici M, et al. Oscillatory mechanics at birth for identifying infants requiring surfactant: a prospective, observational trial. Respir Res. 2021;22(1):314. Epub 2021/12/22. https://doi.org/10.1186/s12931-021-01906-6.
18. Williams EE, Gareth Jones J, McCurnin D, Rudiger M, Nanjundappa M, Greenough A, et al. Functional morphometry: non-invasive estimation of the alveolar surface area in extremely preterm infants. Pediatr Res. 2023. Epub 2023/04/13. https://doi.org/10.1038/s41390-023-02597-z.
19. Williams E, Rudiger M, Arasu A, Greenough A, Dassios T. Non-invasively estimated alveolar surface area in extreme preterms: development and respiratory outcomes at two years of age. Pediatr Res. 2024. Epub 2024/08/26. https://doi.org/10.1038/s41390-024-03527-3.

Correct Answers to Chapter Questions

Chapter 1 The Respiratory System Before and After Birth

Question 1: a, c, and d
Question 2: e
Question 3: c

Chapter 2 Common Neonatal Respiratory Diseases

Question 1: e

Chapter 3 Ventilation

Question 1: e
Question 2: d
Question 3: a
Question 4: e

Chapter 4 Diffusion

Question 1: d
Question 2: b
Question 3: e

Chapter 5 Perfusion

Question 1: c
Question 2: b
Question 3: d

Chapter 6 Ventilation to Perfusion relationships

Question 1: d
Question 2: c
Question 3: a

Chapter 7 Oxygen Transport to the Tissues

Question 1: d
Question 2: c and d
Question 3: b

Chapter 8 Mechanics of Breathing

Question 1: a and d
Question 2: d
Question 3: d
Question 4: c
Question 5: e
Question 6: b and c
Question 7: c

Correct Answers to Chapter Questions 169

Chapter 9 Work of Breathing

Question 1: d
Question 2: b and d
Question 3: b
Question 4: c

Chapter 10 Control of Respiration

Question 1: c

Chapter 11 Waveforms

Question 1: d
Question 2: b
Question 3: c
Question 4: c
Question 5: a
Question 6: b
Question 7: c
Question 8: c

Chapter 12 The Neonatal Respiratory System at Critical Extremes

Question 1: d
Question 2: a
Question 3: e

Chapter 13 Children and Adults with a History of Neonatal Lung Disease

Question 1: c
Question 2: b

Chapter 14 Lung Function Tests in Neonates

Question 1: d

MIX
Papier aus verantwortungsvollen Quellen
Paper from responsible sources
FSC® C105338

If you have any concerns about our products,
you can contact us on
ProductSafety@springernature.com

In case Publisher is established outside the EU,
the EU authorized representative is:
**Springer Nature Customer Service Center GmbH
Europaplatz 3, 69115 Heidelberg, Germany**

Printed by Libri Plureos GmbH
in Hamburg, Germany